John R. Leifchild

On Coal at Home and Abroad

with relation to consumption, cost, demand, and supply and other inquiries of

present interest

John R. Leifchild

On Coal at Home and Abroad
*with relation to consumption, cost, demand, and supply and other inquiries of present
interest*

ISBN/EAN: 9783337100933

Printed in Europe, USA, Canada, Australia, Japan

Cover: Foto ©Andreas Hilbeck / pixelio.de

More available books at **www.hansebooks.com**

ON COAL

AT HOME AND ABROAD

WITH RELATION TO

CONSUMPTION, COST, DEMAND, AND SUPPLY

AND OTHER INQUIRIES OF PRESENT INTEREST

BEING

THREE ARTICLES CONTRIBUTED TO THE EDINBURGH REVIEW

With an Appendix

BY

J. R. LEIFCHILD, M.A.

AUTHOR OF 'OUR COAL AND OUR COAL-PITS; THE PEOPLE IN THEM, AND THE
SCENES AROUND THEM'; 'CORNWALL, ITS MINES AND MINERS, WITH
SKETCHES OF SCENERY'; 'THE HIGHER MINISTRY OF NATURE
VIEWED IN THE LIGHT OF MODERN SCIENCE, AND AS
AN AID TO ADVANCED CHRISTIAN PHILOSOPHY'

. LONDON

LONGMANS, GREEN, AND CO.

1873

PREFACE.

THE PUBLISHERS having determined to republish my recent Article on the Consumption and Cost of Coal as it appears in the current number of the *Edinburgh Review*, I have thought it advisable to add two Articles upon other aspects of Coal and Coal Mining, which had previously appeared in the same Review.

The second Article, which appeared in January, 1860, though nominally a review of the late Professor Rogers' work upon the Coal Fields of America, contains much general information upon the mineral fuel itself; and the third Article, which appeared in April, 1867, although specially devoted to the consideration of Fatal Accidents in Coal Mines, nevertheless includes many details of Coal Mining, particularly as it is carried on in the great North of England Coal Pits, which will help to acquaint the reader with certain conditions of mining exercising a considerable influence upon the extraction and cost of coal.

I have not thought it desirable to alter either the second or third Article, and have only added three notes.

I have written an Appendix, in expansion of one or two topics briefly treated in the first Article, and with a view to put the reader in possession of the latest information and figures respecting special matters of enquiry. The evidence very recently given before the Committee of the House of Commons, now sitting, remarkably corroborates the main positions and statistics advanced in my Article upon the Consumption and Cost of Coal; and some portions of this evidence which throw fuller light upon interesting topics are added in a condensed form.

I likewise venture upon a forecast as to the future cost and supply of coal to ourselves, which will doubtless interest the general public, whose attention has lately been so anxiously directed to this important subject.

J. R. L.

London: *May* 1873.

CONTENTS.

———◦———

COAL

AT HOME AND ABROAD.

—◦◦◦—

I.

CONSUMPTION AND COST OF COAL.[1]

THE WINTER through which we have just passed will long be memorable for the high price of coal, and the privations which this cost has inflicted on the poor and the lower middle classes; nor have any, excepting the rich, been insensible to the inconvenience arising from this cause. With the poor, indeed, it has been a matter of health or disease, of life or death, and only those who have habitually visited them are really aware of what they have suffered, and only in a less degree, are still continuing to suffer, from the lack of fuel. The commonest topic of conversation has been this, and the daily enquiries on all sides were, ' What has occasioned this immensely increased cost; how long is it likely to con-

[1] I. *Report of the Commissioners appointed to inquire into the several Matters relating to Coal in the United Kingdom.* 3 vols. 1871.

II. *Mineral Statistics of the United Kingdom of Great Britain and Ireland for the Year* 1871. By Robert Hunt, F.R.S., Keeper of Mining Records. 1872.

III. *Industrial Partnership, a Remedy for Strikes and Locks-out.* A Lecture delivered at Sheffield, March 9, 1870, by Archibald Briggs. (Reprinted.) 1871.

tinue? who profits by it to the greatest extent? and what
have been and what will be the consequences in trade and
commerce, as well as in household expenditure?' These
are the enquiries which we propose to ourselves, and hope
to answer, as far as the complexity of the whole subject
will enable us to do so in one article; and we shall conclude by suggesting some issues of the gravest national
importance, unless remedies be discovered and adopted
for our present evils.

The cause, or rather the causes, of the present cost of
coal are several, and require to be distinguished one from
another.   There is a general, long-prevailing, and in
some respects a happy cause for the increased cost—
namely, the rapid general extension of our national industries; the return of prosperity after a weary interval
of depression; and a remarkable revival of some particular
industries which require a large supply of coal.   So far,
a great demand for coal is not a subject for lamentation,
but for congratulation; that is, the nation benefits at the
expense and to the inconvenience of numerous individuals
composing it.

This fact is strikingly illustrated by the late very
vigorous vitality of the iron manufacture and iron trade,
which is mainly dependent upon the supply of appropriate
coal.   Ever since the invention of the puddling process
the foundation of iron-making has been coal.   Given a
good supply of coal of a suitable quality, and almost any
kind of iron can be made of marketable value by working
and re-working it.   Hence the quality of iron ore has
become subordinate to the abundance of coal, and the
great centres of the iron manufacture were naturally fixed
on such coal fields as yielded the best and most ready
mineral fuel.   Yorkshire, Staffordshire, South Wales,
Durham, and Northumberland have become the centres
of iron because they are by nature the centres of coal.

It is popularly supposed that the excellence of iron depends on that of the ore, but in truth it is more directly dependent upon the coal; for the amount of work put into the iron is equivalent to the quantity of fuel burnt in producing and re-working it. In Staffordshire about 24 cwt. of coal (long weight, or 120 lbs. to the cwt.) are consumed in producing from pig-iron one ton of puddled iron, the rate of consumption being about four pounds per minute, or 240 lbs. per hour; but in respect to the superior iron, with the most economical mode of working in the present practice of Staffordshire, the making of bars marked as ' best, best, best,' corresponds to a consumption of five tons of coal per ton of iron made from the forge pigs, which themselves require about two tons of coal for their production. None of these quantities, however, are permanent by necessity.[1] Great improvements may, and we earnestly hope will, be realised. One should be instanced. Bessemer's process has saved us half a million of tons of coal in converting 150,000 tons of steel in one year,[2] and it is said that a recent invention

[1] A great saving in the coke and coal used in iron-making has been gradually effected of late years. We gather from Mushet that 5 tons of coal were necessary for a ton of pig-iron in 1810 in Staffordshire, and much more previously in the same country. In Mushet's time nearly 4 tons of coke were needed to produce a ton of pig-iron; the latest information we have shows that at Witton Park, 23 cwts. of coke are required for each ton of forge pig-iron.

[2] It is estimated that in one year, 1869, the consumption of coal for the make of pig-iron was 16,337,271 tons, and the coal used in the conversion of this pig-iron into malleable iron was 15,859,335 tons. The total coal used, therefore, in our iron manufacture in 1869 were 32,207,706 tons. The pig-iron produce in Great Britain during 1871 exceeded six and a half millions of tons—viz. 6,627,179 tons—and then the value of this production at the place where it was made was 16,667,947l. No doubt the iron produce of 1872 was much in excess of that of 1871, so that we may safely assume that at least 7,000,000 of tons of pig-iron were produced in Great Britain, and probably considerably more. Proportionately additional coal was consumed in obtaining this amount, and proportionately more in

of Dr. Siemen's for making steel will be equally effective
in point of economy.

, The above refers only to one branch, though that is
the principal branch, of our metallurgy. If we look to
others we find similar results. In all sorts of copper-
smelting, to take another principal branch, it is calculated
that upon all kinds of ore the consumption of coal is not
less than eighteen tons for every ton of copper produced.
In 1859 Dr. Percy stated that in certain works from 13
to 18 tons of coal (which then cost 5s. per ton) were required to make one ton of copper, and that about half of
this quantity was consumed in the first and second operations of calcining and smelting. Hence copper works
must be the first to suffer from an extraordinary rise in
the price of fuel, which forms the largest item of their
expenditure. Dr. Percy estimates that in works such as
he supposes there would be an annual consumption of
about 20,000 tons of coal, or for every ton of copper
made from a mixture of ores yielding ten per cent. of

preparing the copper, tin, lead, silver, and zinc which have to be added to
the iron to arrive at the total of the make of metals during any year. In
the same year (1871) there were 6,841 puddling furnaces at work in the
kingdom. What, therefore, must have been the annual burning of coal for
nearly 7,000 puddling furnaces, besides that for the forging of pig-iron,
and for all the additional metallic productions simultaneously wrought;
and what will it continue to be every successive year? There is, indeed,
some ground for hope of a small reduction in this particular demand by the
introduction of mechanical methods of puddling iron. Dank's (American)
Furnace has been much vaunted, and three large iron-works have arranged
to introduce it in England; one of those indeed has already found it to
work satisfactorily. We observe that a Belgian Commission of Iron
Manufacturers have discussed and reported very favourably upon the merits
of this invention, and have visited Middlesborough for the purpose of testing its applicableness in Belgium. The Report of this Commission is very
interesting to iron-workers, but we here only notice the question of the
saving of fuel by using it. The saving is not considerable, and the Report
says, ' the consumption (of coal) ought not to be greater in Dank's Furnace
than in others;' in fact the consumption during twenty-four hours represented very nearly the consumption of an ordinary furnace with forced air.

copper, 18 tons of coal. Following out such elements of a general estimate, it was found in 1869 that 149,238 tons of coal were used in the entire smeltings of copper ores in Britain. Very large quantities of copper ores and regulus are brought to Swansea to be smelted from Chili, from the Cape, from Portugal and elsewhere, because it has hitherto cost less to bring the ore to the coal than to send the coal to the ore.

In like manner we may pursue our enquiry into the consumption of coal for the reduction of all of the metallic ores raised by us or sent here. Keeping to the year 1869, and adverting to lead, it was estimated that in smelting and desilverising the 966,868 tons of lead ore then raised, about 145,299 tons of coal were burnt, and that for the ten preceding years the average annual consumption of coal for such work of all kinds must have been 141,694 tons. If the lead ore imported by us in 1869 be added, the estimate of coal consumed must be raised to 177,577 tons. Of zinc, it may be added, that if we include the large imports of that metal, our zinc smelters used, for the whole reduction of the zinc in the same year, as much as 231,176 tons of coal.

The whole matter, then, resolves itself into a question of mineral instead of monetary capital and issue. We have a natural bank of bituminous or carbonaceous fuel, and the capital in that bank is a fixed quantity, while at the same time we have a manufacturing demand upon this natural bank which, in such times as the recent and the present, has amounted to a run. The whole nation, indeed, is running upon its natural bank, and in nature as well as in commerce the results are the same, with this difference, that the natural has not, like the national bank, any artificial system of adjustment. It has no prudential reserves, no raising of the rate of discount, except what manifests itself in the augmented cost of production.

This is in fact the correlative of the Bank of England's
variations of the rate of discount. In the deep subter-
raneous cellars of our bank of coal there is so much bituminous bullion, and no more; it cannot be added to, but
can only be subtracted from; there is no influx, but only
efflux; therefore, if you draw too much by a sudden and
simultaneous rush upon it, the pits' mouths are like the
bank-doors—they must be closed for a time, or at least
the price of the supply will be raised.

We must be prepared to follow out this leading truth
into all the departments of trade and manufacture, and to
find that the coal famine, as it has been expressively
called, has been caused by a strong demand for coal in all
directions. Every householder and every manufacturer
has presented his demand, and all of these cannot be met.
The bituminous bank cannot raise its rate of production,
but it can raise the cost of extraction, and does raise it to
the dismay and disaster of all—even the comparatively
rich. Everywhere gloom exists, and very gloomy appre-
hensions prevail as to what is to come.[1] Iron has not
only of late advanced very greatly, but it is still rapidly
advancing. Look at an ironmaker's circular, and you
will read the same kind of intelligence as this, dated
February 14 ult., from the representative of the great
firm, in the following words:—' I beg to inform you that
the price of Earl Dudley's iron this day has been advanced
1*l.* per ton.' Of course other firms will follow their

[1] The intelligence received from many quarters informs us of the partial
or total close of several iron-works. How can they be kept in operation
when the cost of smelting the ore, owing to the cost of coal, exceeds the
profit derivable from the process? It is impossible to work at a loss, and
the question is will that loss be realised. The price of iron may be
raised to a point at which foreign consumers cannot take it; and when it
cannot be sold to advantage, the principal part of this manufacture will be
at an end in our country. (See our note at the close of this article, on
American diminished demands.)

leader. Bars are dearer than they were a year ago by 2*l*., and yet the 12*l*. of the preceding February became 16*l*. in July of last year. Advances up to March 14 of this year have raised the best marked bars of Staffordshire throughout to 15*l*. and 16*l*. per ton. Nor are these advances quite the largest in iron-making; for in another description of the made metal the advance has been 40*s*. within five weeks; if the advance within five weeks be 40*s*., what will it be in five months? Moreover, during the five months including and following after February, the demand is always the most urgent.

Already the consequence of the increased price of iron has told severely upon all undertakings largely requiring it. Our railways have suffered greatly, and the directors of all of them complained loudly at their last half-yearly meetings of this, the principal cause of diminished dividends. The recent report of the London and South-Western Railway, while it gives a very favourable account of this company's affairs, mentions that the increased price of coal, from 15*s*. 3*d*. a ton in 1871 to 22*s*. 10*d*. at present, has, in their requirements for locomotives and traffic purposes, cost no less than 22,000*l*. in the accounts for the half-year. In the last statement of the North Staffordshire Railway Company we read, that whereas the coal for their locomotives had previously cost them from 8*s*. to 10*s*. per ton (they drawing it from their own coal field), coal has recently cost them from 18*s*. to 20*s*., and is still advancing. Colonial Railways have suffered like our own. The Great Indian Peninsular Railway in 1867 paid 51*s*. per ton for coal; for coke, 62*s*. a ton; and for patent fuel, 54*s*. a ton. Nor are these the highest figures. The Madras Lines of Railway, as well as the Railway Companies in Western India generally, are dependent for coal upon England and share in its surcharges for their fuel. Next to iron, gas-works probably

consume one-ninth of all the coal raised, and their consumption is in direct proportion to population and prosperity. Every addition to the habitations of the metropolis and large provincial cities and towns, must be ac-
companied with gaslighting in streets and largely in
houses. Gasometers appear in all directions, and gas
companies will abound and thrive, until they also feel the
cost of coal, and endeavour to raise the price of gas, from
doing which many of them are by their Acts of Parliament restricted; but ultimately they must follow the
general and imperious law of higher prices or smaller
profits to their shareholders.

Our cotton manufactures require every year at least
two and a half millions of tons of coal, and our woollen
and worsted manufactures about one million and a
quarter. In all the manufactures of textile fabrics, the
disadvantage of a rise of price has been similar if not
equal in amount. Manufacturers have declared that the
high price of coal has been to them equivalent to an increase of from one halfpenny to one penny halfpenny in
every pound of cotton. Curiously too, in the Staffordshire Potteries situated in one coal field, it has lately been
found desirable to import coals from another.

In whatever direction we look, nothing can at present
be discerned but a constantly augmenting demand upon
our coal fields. Steam power now makes a very large
demand upon the extraction of coal, and we may warrant-
ably compute it at between twenty-five and twenty-six
millions of tons every year for the United Kingdom.
When we add steam navigation to this, we can scarcely
be in excess if we give in round numbers the large
amount of thirty millions of tons of coal as the total
annual requisite for all steam purposes in manufactures
and navigation. Extended steam navigation, steam
machinery, steam power appearing in numerous new

modes; heating, lighting, cooking, the fashioning of most articles of luxury as well as of necessity, all depend upon coal as the prime motive power.

In addition to the reasons for a rise already stated, another strongly operative cause is the immense and rapid increase of population. It has been computed that for every additional person born an additional ton of coal is required. We take this as an element of average, but if we examine the total quantities of coal annually raised in Great Britain from 1855 to 1870, we shall see that by a table which the Commissioners give (vol. iii. p. 178), the consumption of coal in relation to each head of the population ranges from two to three tons. This of course is a fallacious view, since the demand for manufactures and metallurgy is included in the calculation. Since, however, the domestic consumption of coal in 1869 was estimated at 18,481,572 tons out of the total extraction of 107,427,557 tons, we may assume the domestic consumption at present to amount in round numbers to twenty millions of tons; hence, too, we may fairly assume that an increased supply of at least one million tons is required every five years from the mere increase of population. The effect of this increase in the metropolis and its immediate suburbs is obvious. In every six minutes a child is born in London and its boundaries; hence in every six minutes an additional ton of coal is required, although there is a partial compensation by simultaneous deaths. The increase of London during the ten years from 1851 to 1861 showed that the population will double itself in forty years. London in sixty years of the present century has trebled its inhabitants. At the rate of doubling them in forty years, the number of inhabitants in London in the year 1901 may rise to 5,700,000 human beings. Some Londoners who read these pages may live to hear that there are six millions of fellow-creatures around

them, each one of whom may need a ton of coals. If
the existing three and a half millions cannot obtain coals
enough at a moderate cost, what will be the case of their
successors, and, indeed, their future metropolitan contem-
poraries? Add to this the entire future population of
the United Kingdom, and the anticipation becomes
appalling. Making all due abatement for the uncer-
tainty of statistical facts and deductions, *appalling* is not
too strong a term to apply to even a cautious anticipation. Conceive six or ten millions of Londoners and
Suburbans making the same run upon the national bituminous bank as we are now making—for mineral fuel in
domestic heating and cooking, for street and house gas,
and for various manufacturing purposes, while the same
or similar restrictions of supply prevail—conceive that
this issue may be realised, and then who will predict the
result? Coal has become as necessary to social life as
food is to man. But if population really increased, as
Mr. Malthus supposed, in a geometrical ratio, whilst the
deposit of coal is not capable of any increase at all, it is
evident that at no very distant period, it would be impossible to obtain coal enough for all the wants of society
—at least in this island.

From the continual operation of these combined causes
the reader will be prepared to credit the astonishing
progress of coal extraction in the last few years. If we
begin with the sixty-five millions of tons extracted in
1857, and pass to the seventy-two (nearly) millions of
tons extracted in the year 1859, thence proceeding to
the ninety-eight millions of tons in 1865, we may advance
at once to the one hundred and seventeen millions of
tons raised in 1871 (the latest authentic return).[1] It is

---

[1] See Appendix, on the Output of Coal in the year 1872, where it is
shown that the total output for the United Kingdom in that year was
123,546,758 tons. This fact was unknown at the time the article was
written.

therefore manifest that we have increased our coal extraction by about fifty-two-millions of tons in fifteen years, and that the increased extraction during that period approximates to the total annual extraction of the first year. Furthermore, it seems highly probable that under present causes our total coal extraction will in five years hence be at least one hundred and thirty millions of tons for the year, in which case the entire coal production of Britain will have doubled itself within twenty years. It was said in 1865 that the rate of growth in that period in the aggregate annual consumption of coal, reckoning each annual percentage on the previous year's consumption, amounted to three and a half per cent. per annum.[1] For many years the consumption of coal has really been increasing at the rate of about four per cent. per annum, computed in the manner of compound interest; so that in eighteen years our present consumption would be doubled, and in thirty-six years would be quadrupled, while in fifty-four years it would be eight times more than it now is. Be this as it may, as to its entire truth, the probability previously mentioned is exceedingly strong, viz., that if prices remained the same the present consumption would again double itself in the next twenty years; and this startling conclusion prompts us to ask—whence are we to derive, and at what charges can we deliver, the amazing amount of two hundred and thirty-four million tons of coal in one year? A year or two, more or less, before such a total extraction is realised is of little moment, for we are now in the midst and under the pressure of the consequences, and we have at best but a brief space of time left us in which we can

---

[1] So stated by Mr. Jevons in his work 'On the Coal Question,' in which he anticipated that in 1871 the consumption would amount to nearly 118,000,000 of tons. This is a very close approximation to Mr. Hunt's official return of 117,352,028 tons in 1871.

endeavour to diminish their severity, or to defer their aggravation. If we cannot succeed in so doing, no light calamity is impending over us of the present generation, and a much heavier calamity upon our successors in this country, and perhaps in other countries. If the Bank of England were to break, the whole world would feel the monetary shock : and if the bank of British coal should fail, or approach to failure, it is certain that while many of our envious neighbours and remote competitors would rejoice, they themselves, with others, would participate in the evil effects of such a failure. A bankruptcy of British coal would shake the prosperity of all civilised Europe, as we shall now show that several other countries depend upon us for the coal we raise and they import.[1]

Not only are we called upon to meet the wants of our own busy land, but several countries have made considerable calls upon our coal resources, and are continually increasing their demands, insomuch that our present exports of coal are nearly four times as large as they were twenty years ago.

France is the largest foreign consumer of our coal, and the gradual growth of the exports to that country is truly remarkable. In 1812 we gave France a very small quantity of coal. In 1822, however, we sent there 31,000 tons; in 1832 we exported as much as 37,000 tons ; and in 1842 no less than 490,000 tons. Advancing to 1852, the birth year of the Second Empire, France obtained from us 652,000 tons. In 1862 it was found

---

[1] Were our space ample we should enter into the consideration of possible retarding elements of demand. Mr. Price Williams' elaborate table may be consulted in the Commissioners' Report (vol. i. p. xvi.). It is constructed upon the basis of ratios diminishing according to certain views of his own. According to this table the *annual* consumption of coal at the end of another 100 years would be 274,000,000 of tons ; and further, the *total* estimated quantity of coal available for use would be exhausted by consumption in 360 years.

that the growth of manufacturing industry was so considerable that it had enlarged the coal demand of France upon us to 1,306,255 tons; while, in 1872, it rose to 2,191,340 tons. Thus steadily have our exports of coal to France grown from a few to many thousands of tons, and then to millions, so that the total increase in the fifty years, from 1822 to 1872, has been 2,160,235 tons. We find that the present total annual extraction of coal in France itself may be estimated at 14,000,000 tons, and therefore it appears that we send to it more than one-seventh of its own coal produce. In fact most of the same causes have there effected an increased consumption of coal as in Great Britain. Augmented steam power has led to augmented consumption of steam coal; and as the French have in all likelihood doubled their railway mileage since 1853, so more coal has been wanted for locomotives. In nearly all the manufactures and trades in which we have prospered they have prospered in the same proportion, and coal has become as essential to them as to us. They also have wonderfully increased their own coal extraction, so that they can compare their present annual 14,000,000 tons with a mere extraction of 770,000 tons in 1813; and they also will continually raise more coal themselves, and want more of our coal from us. They now take about one-sixth of the total shipped by us in exportation.

Of Germany, with some differences of detail, nearly the same might be said. That country does not take quite as much coal as France, but it also will be continually helping to drain us. South America took one million of tons, and Russia three quarters of a million of tons of coal in 1872. The total amount of coal shipped by us last year to foreign countries was 12,092,000 tons, showing an increase of 302,027 tons over the shipments of the previous year; and every year our exports have been increasing, although it was thought in 1869 that we had surely arrived at a maximum when we shipped nearly ten and a half millions of tons, and, including coke and anthracite, actually 10,837,804 tons.

While existing commercial treaties last we cannot impose a duty on the export of coal, and it is against our national policy to prohibit the exportation, even if we had the power so to do. But there are still more decisive reasons against resorting to restrictive measures of this nature. They would tell in the most fatal manner against ourselves. A large portion of the coal we export is for the service of our own steamers in all parts of the world, and a great number of British ships are engaged in the foreign coal trade. Hence it may be inferred that a heavy export duty on coal would be highly injurious to our maritime interests, and would be in truth a disastrous tax on steam navigation abroad and on freight.

So far we have dealt mainly with the quantities of coal extracted, and the rapid increase of this extraction; but we are now brought to the consideration of prices— the *cost* of *coal* as well as its consumption; and here we have a fluctuating instead of an actual constantly advancing and calculable element. We should be glad indeed to give this element fuller consideration than our space permits, because it tends to govern consumption, checking it when high and enlarging it when low.

It is curious to examine the fluctuating prices of coal at a remote date and downwards to our day. So long ago as the year 1635 coals cost in London 10s. per London chaldron, the lowest price to which we can trace them, as well as the earliest date. In 1665 they had risen to 13s., in 1761 to 24s. 9d., in 1768 to 36s., in 1785 to 40s., and in 1793 to 42s. 6d. per London chaldron in London. In 1805 we find them at 44s. 9d., in 1810 at 51s. 8d., in 1819 at 59s. 1d. at the ship side in the port of London per London chaldron of 25 cwts., which was rather less than half of the Newcastle chaldron of 53 cwts. From this culminating price there followed a descent in subsequent years, down to 33*s.* 6*d.* in 1831. In 1832 the average price was 21*s.* 11*d.* per ton, from which a rise ensued to 23*s.* 8*d.* in 1839, and thence again a varying descent down to 20*s.* 2*d.* per ton, always at the ship side, and apart from duty. From all these details it seems that the price of 10*s.* per chaldron in 1635 became doubled in 1761, and then again became more than doubled in 1793, when the price was 42*s.* 6*d.* per chaldron. There were, therefore, in old times causes in full operation which doubled and quadrupled the relative prices of coal, though we cannot now ascertain their precise nature; but we thus see how similar and fluctuating causes have in like manner sometimes doubled, and, as at present, even more than doubled, the cost of coals, to London consumers at least, who must deal with London coal merchants. Persons not paupers, but in the condition of economical lodgers, have this last winter been paying as much as five shillings for a single sack of coals of inferior quality.

As the recent very high cost is now a special subject of enquiry by a Parliamentary Committee, it is needless here to dwell upon it, much as it has been talked and written about in all circles. Nobody, indeed, at present really knows where the chief blame lies, and to whom the chief gain accrues. We have conversed with merchants on the London Coal Exchange, with coal owners, with subordinate dealers, and with miners, but without any decided and clear result. Mutual recriminations are the fashion, and each class flatly denies the affirmations of the other. They must all be examined and confronted in order to elicit the truth.

Some principal elements of the enquiry respecting recent

and present cost are tolerably definite, and may be fairly adduced. They are such as these: an extraordinary demand occasioned by revived iron-making and manufacture has lately prevailed, so that for a year or two vastly more iron-working has been in operation than for several previous years. A season of activity has succeeded one of prostration, and consequently a largely increased demand for coal has ensued, and raised its price. Many other leading manufactures have in a greater or less degree simultaneously revived, and have in a greater or less measure demanded more coal. Coal miners well knew this, and naturally inferred that the coal owners were making immense profits, of which they, the workers, ought to have a share; they struck for such participation in many coal fields, and made their wages rise by forty and fifty and in some few instances one hundred per cent. in little more than one year. The pitmen thought that they had the key of the situation, as indeed they temporarily had. Dear pit labour is dear coal; dear coal is dear iron; and dear metallic products affect commerce in general. Hence in winter we have cold comfort, and a cold home. Diminish colliers' wages, if you can, and you get, at least as many think, at the root of the matter. The pitmen advanced in their demands upon the owners or proprietors of pits, proportionately as they thought the owners were advancing upon the public consumer. The owners had too long been getting the lion's share, and the pitmen would have their fair share—being assured that their masters were gaining to a disproportionate extent. Upon this, the rise in pitmen's wages was made the ground of a further advance in the price of the commodity.

Circumstances have lately brought matters to a crisis, and while the two classes were contending, of course the public had to pay, and to pay inordinately. On their side the pitmen have clever calculators with many figures, and they fill their journals with strong assertions. The difficulty is to discover the truth. Newspapers in all directions have published notes of the price of coal at the pit's mouth, and in London coal offices. The difference was so great that some middleman or middlemen must be public plunderers. Coals selling eighteen months ago, as Mr. Mundella avouched, at from 18s. to 20s. per ton are now, or recently have been, selling at from 45s. to 50s. per ton, while the miners' wages, which a year and a half ago were 2s. 6d. per ton, are now only 3s. 2½d. per ton. Were this a persistent condition the rise rests either with the coal owners or coal merchants, or upon both in combination.

Now let us listen on the other side to a coal owner in conference with his miners and others, one of the few who has consented to explain as well as to complain that he lay under unjust odium. At the Clifton Collieries, near Nottingham, the new proprietor lately asserted that it could be shown by the books that out of the increased prices current labour now received a very much larger proportion of the increase than the proprietary. He entered into details and averred that in 1870 the 'stall-men' (the workers in the *stalls*, or mining galleries), received 2s. 3d. for getting a ton of coal, and the wages paid to the unskilled labourer were from 3s. to 4s. 6d. a-day. Now the stall-men receive 4s. 3d. for getting a ton, equal to 95 per cent. of increase on their previous daily pay, while the day-labourer can earn 6s. 6d. a-day. Considering the large addition of outlay required by the almost doubled price of timber and rails, the proprietor contended that the labourer got a larger proportion of the increased price of coals than he himself did, although the realised price of coals is now 12s. 4d. a ton, whereas in 1870 it was but 6s. 9d. per ton. He, the proprietor, was now receiving, as the capitalist, on this excessive rise

c

a profit of 17½ per cent. upon his sales, while labour is
actually realising 75 per cent. additional as compared
with the wages received in 1870.   Such is the best
exculpation, upon an owner's part, within our reach.

In continual perplexity, we turn again to the pitmen,
and ask, ' What is your comment upon this explanation ? '
' A denial of the statement,' they reply. The best method
of illustrating this denial is to proceed to the great coal
fields of the North of England, and to take the facts
there offered to us on the side of the miners, by a local
paper.   In these great coal fields the owners are affirmed
to have been realising about 13s. per ton on the prices of
coal maintained for a year past, and if so their gains must
be inordinate ; for the thirty millions of tons of coal now
annually raised in Northumberland and Durham would,
at 13s. per ton of profit, yield a clear revenue of more
than nineteen, or nearly twenty, millions of pounds.
Amongst how many owners is this distributed ? It is said
that the large owners in the two northern counties number about two hundred, each of whom, therefore, would
realise nearly as much as 100,000l. in one year ; a
princely income, indeed, and a liberal reward for the
exercise of ' an enlightened self-interest ' !

There is a third party, or a third plunderer in public
estimation, whom we have in passing to consider—the
great London coal merchants.   Their name is not
' Legion,' for they are said to number hardly twenty ;
but their alleged misdeeds are very suggestive of an
unholy alliance, and certainly they have plagued us very
sorely, and could not be cast out.   It has been said that
they have kept back coal, and artificially raised its price,
and that they have ' rigged ' the London coal market as
stockbrokers often ' rig ' the Stock Exchange.   It is not
our duty to examine, or charge, or exculpate them, only,
till they are publicly and clearly exculpated, they must

expect to bear a bad name. Unluckily for them appearances are against-them, and they must make their case much clearer than it is, if they would escape from general obloquy.

The recent diminished output of coal from our principal coal fields is a consequence not a cause, or, at all events, only a secondary cause, of high prices. Here we have only to establish the fact that the output has of late been considerably less, and we shall be led to the conclusion that by contrivance and collusion it can at any similar conjuncture of conditions be brought about in like manner, if not in like measure.

From Durham the output in 1872 was less than it was in 1871 by 350,000 tons. In Lancashire the amount of coal raised in January 1873 was from three to fifteen per cent. below the average, and in some other districts about twenty-five per cent. below the average brought to bank.[1] It is very difficult in such disturbance of the equilibrium of a great trade to trace and assign to each disturbing element its prime value. This it will be the duty of the Select Committee recently appointed to enquire into this subject to accomplish.

---

[1] If we compare the chief railway deliveries of coal from the collieries in the chief coal fields of our country for two consecutive months, viz., December of last year and January of the present year, we discover a singular falling off in the amounts carried to London in January, although that was by far the colder month. Taking some examples in the Yorkshire coal fields, the Great Northern Railway carried only 65,125 tons in January as against 91,181 in December. Of the Silkstone coals, which have been much used in London for household purposes, 9,248 tons were brought in January against 14,319 in December; and of the Barnsley thick coal only 11,509 tons were carried in January as against 15,677 in December.

By the Midland Railway, which serves the Derbyshire coal fields, there arrived 80,775 tons of coal in January, but in December 90,511 tons. The same contrast might be drawn in relation to other coal fields; and from this it is manifest that by so much less coal than ordinary was delivered to Londoners, and London coal merchants were compelled to charge more unless they themselves conspired to keep back the delivery.

We say little respecting existing or previous strikes,
because these are phenomena of passing times, often fully
detailed in the journals of the day, and happily forgotten
when past, though unhappily the lessons they convey are
equally forgotten.   If the terrible sufferings produced
by every great strike were remembered, and if the results
of experience were recorded and acted upon, such strikes
would seldom, if ever, recur.   Existing strikes may slowly
or suddenly cease, and presently discordant elements will
disappear, but the permanent and progressive causes of
demand which we. have been anxious to elucidate, will
continue, unless we can institute radical and national
economies.   In all our observations we have more regard
to these lasting economies than to temporary questions of
cost, however urgent.   In all such considerations we have
to look at great natural and commercial conditions, to the
countervailing evils which follow an excessive national
prosperity, and to the fact that our national prosperity is
founded upon coal and our manner and rate of using
it.   The misfortune is that our knowledge grows only with
our exigencies, and that our exigencies alone quicken our
desire for knowledge.   Most of the information that the
great public have gained about coal has reached them
very late, and not only late but also in irregular instal-
ments.   What has been buried in 'blue books' ought to
have been published on the housetops.

It is an elementary proposition in political economy
that a rise of prices tends to correct itself by a twofold
influence: it increases supply by attracting more labour
to a profitable field, and it restricts demand by reason of
the increased cost of the commodity.   But in this case of
coal, these general principles have been counteracted by
some peculiar and exceptional causes.   The rise in prices
and in wages, instead of bringing more labour into the
market, diminished the actual quantity of work done;

because the colliers, finding they earned as much by three days' labour as they had earned before in five days, preferred to take more leisure rather than more money. Although wages had advanced 40 per cent., the total amount earned in many of the collieries was positively less than it had been before the rise. Of course the 'output' of the pits diminished in the same proportion; and the colliers found that the less they worked the more highly they were paid. This depended on their having a virtual monopoly of labour in the pits, owing to the difficulty of training new men to so laborious and repulsive an employment, and to the opposition the men would themselves offer to any new comers. The only true remedy of the evil is, in our opinion, to be found in absolute freedom of labour. The men have a right to refuse to work more than three days a week, if such is their pleasure; but they have no right to combine against the introduction of independent labour from other quarters. To leave the pitmen in possession of an exclusive right to work the collieries would be to leave the fate of the nation in their hands. Other underground men can be found, or machinery can be introduced. It is impossible to admit that the price of coal is to be permanently raised by a monopoly of the labour which extracts it.

Freedom of labour may restore the supply; to limit the demand we must look to economy: and in this view of the case, we are not sorry that the people of England should have had a severe lesson on the clumsiness and extravagance of their arrangements for producing heat. The extreme abundance and cheapness of coal had made this country the worst provided in the world in its domestic arrangements for warming and cooking. The same amount of warmth may be obtained at one-fifth the expense of fuel; and the huge British kitchen range is absolutely destructive of all good cookery. Recent im-

provements in steam boilers have demonstrated that an
equal economy is quite practicable in the production of
steam power.   Henceforth the profits of trade will be
found to depend to a considerable extent on a strict
economy of fuel.

The Commissioners whose Reports we have placed at
the head of this article are half a century too late, and
diligent as they have been, in this case the hand of the
diligent has not made us rich, but displayed our poverty.
The emergency in which we are now placed was not and
perhaps could not have been anticipated by them, and so
we really mourn with a justifiable sorrow as we turn over
pages after pages of details which have little present or
practical utility.   On some main points, however, they do
afford us valuable information.

When we arrive at an examination of the evidence for
the total amount of coal probably remaining to us in this
country we have first to estimate the total, and then to
draw a line between the available and the inaccessible
coal.   There is in all geological probability a vast quantity of this mineral fuel buried at inaccessible depths.
Taking for the moment the extreme limit of accessibility
at 4,000 vertical feet, there are beneath that limit, according to well-founded conjectures as to what lies below and
between the Permian and other newer strata, about
41,144 millions of tons of coal.   This last total is composed of more than 29,341 millions lying at depths varying from 4,000 to 6,000 feet, while 15,302 millions might
be found at depths of from 6,000 to 10,000 feet.   In
respect of the temperature of the earth, that would be
150° Fahr. at a depth of 6,000 feet, while it would be
215° Fahr. at a depth of 10,000 feet, that is three degree
above the temperature of boiling water at the sea-level.

Within the area of known coal fields, about 7,320 millions of tons of the mineral fuel lie at greater depths than

4,000 feet; and of this quantity probably 5,922 millions of tons rest between the limits of 4,000 and 6,000 feet in depth, and the remaining 1,397 millions of tons at between 6,000 and 10,000 feet. Hence the combined totals of all coal conjectured to lie at greater depths than 4,000 feet is a little more than 48,465 millions of tons. If our posterity can in any way contrive to reach and extract any of this deep coal, by so much will they be the warmer and the wealthier, and in order to become both they must necessarily be also wiser in economy than ourselves. They may, however, be cleverer and yet be colder.

Our own concern is restricted to shallower coal; and the Commissioners, after instituting careful and extended enquiries, inform us that the probable quantity of coal contained in the ascertained coal fields of the United Kingdom is 90,207 millions of tons, while the quantity which probably exists at workable depths under the Permian, New Red Sandstone, and other superincumbent strata in our kingdom is 56,273 millions of tons; together forming an aggregate of 146,480 millions of tons of coal, which may be reasonably expected to be available for future use from the time of enquiry.

The essence of all the researches and conjectures as to the probable duration of the above-named quantity may be given in a few lines. Basing these estimates upon present consumption (it is important to distinguish this from *increasing* consumption), the relation which the total amount of 146,480 millions of tons bears to the consumption of 115 millions of tons (in 1871) is as follows: The available total just stated will support our production as at present for 1,273 years; the same quantity would support an annual production of 146 millions of tons for one thousand years; and one of 175 millions for eight hundred and thirty-seven years. Doubling the recent annual consumption, that is making

it 230 millions of tons, this would be supported by the
estimated supply for six hundred and thirty-six years.

Such is the shortest and most popular mode of stating
the results of a long and laborious investigation bearing
upon the future. The concluding expressions of the nine
competent Commissioners in their General Report on
this enquiry are well worth quotation.

Whatever view may be taken of the question of the duration
of coal, the results will be subject to contingencies, which cannot
in any degree be foreseen. On the one hand, the rate of consumption may be thrown back to any extent by adverse causes
affecting our national prosperity; and on the other hand, new
discoveries and developments in new directions may arise to
produce a contrary effect upon the consumption of coal. Every
hypothesis must be speculative; but it is certain that if the
present rate of increase in the consumption of coal be indefinitely
continued, even in an approximate degree, the progress towards
the exhaustion of our coal will be very rapid.

In all the foregoing estimates of duration we have, for the
sake of simplicity, excluded from view the impossibility of supposing that the production of coal could continue in full operation
until the last remnant was used, and then suddenly cease. In
reality a period of scarcity and dearness would first be reached.
This would diminish consumption and prolong duration; but
only by checking the prosperity of the country.

The *absolute* exhaustion of coal is a stage which will probably
never be reached. In the natural order of events the best and
most accessible coal is that which is the first to be worked, and
nearly all the coal which has hitherto been raised in this country
has been taken from the most valuable seams, many of which
have in consequence suffered great diminution. Vast deposits of
excellent and highly available coal still remain, but a preference
will continue to be given to the best and the cheapest beds; and
as we approach exhaustion the country will, by slow degrees,
lose the advantageous position it now enjoys in regard to its coal
supply. Much of the coal included in the returns could never
be worked except under conditions of scarcity and high price.

'A time must even be anticipated when it will be more economical to import part of our coal than to raise the whole of it from our residual coal-beds; and before complete exhaustion is reached, the importation of coal will become the rule, and not the exception, of our practice. Other countries would undoubtedly be in a position to supply our deficiencies, for North America alone possesses tracts of coal-bearing strata as yet almost untouched of seventy times the area of our own. But it may be doubted whether the manufacturing supremacy of this kingdom can be maintained after the importation of coal has become a necessity.[1]

Respecting the possibility of working at great depths below the surface, the Commissioners took great pains to obtain evidence, and to form sound opinions. They printed a series of seventy-nine questions, and distributed 530 sets of these in circulars, to which the replies appear to have been exceedingly disproportionate, although the *vivâ voce* evidence which they obtained was sufficient. In brief, their conclusions are as follows:—The workable depth of coal mines depends upon human endurance of high temperatures and the possibility of reducing the temperature of the air in contact with heated strata. The mechanical difficulties connected with increased depth, and the cost of steam power for hoisting the deeper coal, do not appear too formidable; while the extra pumping of water is met by the presumption, that water is seldom met with in mining for coal at great depths, nor, as a rule, are deep mines more liable to inflammable gas than shallower mines.

The increase of temperature, then, is one main point of consideration. In this country the earth's temperature is constant at a depth of about 50 feet, where the temperature is 50° Fahr. The rate of increase of temperature is in our coal mines generally 1° Fahr. for every 60 feet of depth. It is questionable, however, whether after a great depth the rate of increase does not prove more

[1] *General Report*, vol. i. pp. xvii—xviii.

rapid than before. The best test we have is that of the
deepest coal pit in Great Britain, viz., that at Rosebridge
near Wigan, where the shaft is now 2,376 feet deep, and
is still descending lower and lower. There the ratio of
heat-increase agreed with the ordinary rate down to a
depth of 1,800 feet, after which it became considerably
more rapid. At the lowest point of the sinking the
thermometer indicated 92° Fahr.

Much more is said about temperature, and its equality
and diversity, but the few foregoing and following facts
are enough to enable us to understand the conclusions in
relation to it. What is the maximum temperature of air
compatible with the healthful exercise of human mining
labour? Now the normal heat of our blood is 98°, and
fever heat commences at 100°, and the extreme limit of
fever heat may be taken at 112°. Dr. Thudicum, a
physician who has specially investigated this subject, has
concluded from experiments on his own body at high
temperatures, that at a heat of 140° no work whatever
could be carried on, and that at a temperature of from
130° to 140° only a very small amount of labour, and
that at short periods, was practicable; and further, that
human labour, daily and during ordinary periods, is
limited by 100° of temperature as a fixed point, and then
the air must be dry; for in moist air he did not think
men could endure ordinary labour at a temperature
exceeding 90°. Dr. Sanderson added useful testimony
in detail leading to a similar conclusion, observing that
gymnastic exercises can be practised by men in high
temperatures up to a certain point, but that immediately
when the temperature of the body rises to 102° or 103°
Fahr., then all capacity for further exertion ceases. A
case in Cornwall was instanced of the excavation of
mining galleries where the air was heated by a hot spring
to a temperature said to amount to 117°. Dr. Sanderson

visited this mine, and found the highest temperature to be 114½° Fahr., and the total duration of each of the men's work who were there engaged was less than three hours in the twenty-four. When urged to express the limit of temperature which he considered consistent with continuous healthy labour during five hours at a time, Dr. Sanderson replied, 90° Fahr., with the observation, that a man could not or would not do as much work in moist air at 90° as he could in ordinary conditions; and even at 90° the loss of working power would be very considerable.

The temperature of the earth at 3,000 feet deep would probably be 98° in England. Under what is technically called the long-wall system of working the coal, a difference of about 7° appears to exist between the temperature of the air and that of the strata at the working faces, and this difference increases 4 per cent. a further depth of 420 feet; so that the depth at which the temperature of the air would become, under present conditions, equal to the heat of the blood, would be about 3,420 feet. As to depths beyond this the Commission declined to speculate, but they thought that the ultimate limit of coal-working could be reached. Still many important details in the evidence on this question would, it appears to us, have to be reconsidered in all such deep coal-mining.

Besides the physical capacity of human endurance and existence at any such depths, the increased cost of working and winding up the coals, the greater wear and tear of materials as well as men, and the augmented difficulties of penetration and extraction, and of propping up roofs, would have to be considered, and would all tend to enhance the cost of coal, until perhaps such increased cost would bear such a large proportion to increased depth as to cause the financial to equal or exceed the mechanical obstacles. The deeper the pits

the larger the initial cost and the greater all subsequent
expenses. If to win coal lying at a depth of, say 2,000
feet, costs 100,000*l.*, to win coal at 4,000 feet might
require 250,000*l.* Add to this that the rise in the cost
of all mining materials has been as great as in other com-
modities, and we foresee limits financial and limits mechanical both combining against us; and where human free
will, or rather ill will, superadds its opposing combination,
it would hardly help us if half our earth were composed
of coal or down to its centre while we could not use it.

' Waste not, want not,' is a proverb as applicable to
coal as in common life. We *have* wasted coal and
therefore we do want it, and we have wasted in several
ways, and to a most lamentable extent. Every one
acquainted with coal-mining knows how much of this
invaluable fuel has been absolutely and for ever lost by
bad methods of working. It may be affirmed that during
many years the amount of coal wasted by leaving pillars
needlessly large to support the roof, by clumsy and quite
unscientific methods of getting the coal, and by rough
modes of carrying and delivering it, has amounted to fifty
per cent. of the total; that is, that fully one-half as much
coal has been wasted as has been delivered to the consumer.

It is melancholy also to learn that in what is termed
the ' waste ' and the ' goaves ' of many large coal pits,
some of which have been shut for ever, thousands upon
thousands of tons of the best coal lie buried as in a
fathomless sepulchre. Improvements in mining in the
North of England have allowed of a much less wasteful
extraction there than previously; but taking all our
coal fields together, the ordinary and unavoidable waste
amounts to at least ten per cent. of their whole delivery,
while the avoidable waste sometimes reaches thirty or
forty per cent.[1]

[1] One singular example of waste is that of the ' pit-heaps,' which are
known to colliers. We ourselves were wont to look at these vast mounds of

Early miners in newly explored districts will naturally perform their work in a primitive manner, and hence we are not surprised to hear that similar wastefulness characterises the working of the coal fields of other countries. An American authority estimates the entire waste in the mining of the anthracite of Pennsylvania as fifty per cent. of the total extracted. This loss of mineral seems the more inexcusable as the greater part of the present delivery of anthracite is extracted from a single bed called the Mammoth Seam. That seam is now exhausted above water level, and is known to depreciate below. If the extraction should increase like our own, and augment, as it is said now to do, by about five per cent. per annum, and should double itself in twenty years, it is easy to foresee the imminent exhaustion of that immense coal seam at available depths.

Out of all the coal which we have been burning for centuries nothing is surer than this, that we have never obtained a quarter of its theoretical heating value. We have squandered our mineral fuel like prodigals, with no better excuse than that we were in part helpless in our prodigality. As we know that our steam boilers now consume scarcely half as much coal as they consumed ten years ago, and as the present calorific effect is only one-eighth of the coal actually consumed, what must have been the waste of coal retrospectively for many years! In fact we may be said to have been burning coal in

small coal, which had by annual accumulations swelled into very considerable mounds, and to wonder at the fearful waste of fuel therein involved. Many years ago we stood upon an eminence at South Hetton, and looked over a vast area of these pit-heaps, which, in some instances, were burning away during the night. All the colliers had free access to these accumulations of small coal, and filled their scuttles as often as they pleased. Now, however, the just retribution has arrived. These neglected pit-heaps have become valuable and chargeable; and what had been recklessly wasted for half a century is now sought with money and the remnants are sold to eager purchasers.

systematic waste; even now our knowledge of the laws of heat, and the adaptations of mechanism, do not combine in our favour as we should expect from the rapid advances of practical science. Other countries have been and are more economical in the consumption of coal in their boilers, and this, as well as the great cost of the fuel, should stimulate the inventiveness of our mechanical engineers. Some of them think that we cannot confidently anticipate immediate economy of fuel. They argue, that since we cannot transfer above one-eighth of the total into mechanical power, while the natural conditions remain the same, and the same materials are acted upon, we must not expect a future economy of more than two-sevenths, and that even this economy will probably only be effected in the next generation.

As the consumption of coal in iron works of all kinds consists of about one-fourth of our whole extraction, it would be highly desirable if a considerable economy could be effected in this department of industry. But we have little immediate hope of its realisation. We have said that Dank's rotary furnace is not expected to save much fuel, nor do other hoped-for improvements promise much more immediately. In other branches of metallurgy there is similar waste of coal, but there is little prospect of saving excepting in copper works, where economy is plainly possible.

Nearly every householder has been lately discussing, and often adopting, expedients for economising his costly fuel, and much has been written and said about grates and stoves, and bricks and iron plates, and fire-balls. Here again we want because we have wasted; we have all been using open grates, and these probably deliver to our apartments an amount of heat which may be represented as one-twentieth of the total heat capable of being extracted from the fuel they consume. When, however, we are advised to throw forward our domestic grates, we have only a very partial remedy proposed to us. Throw them as far forward into our rooms as we will, the heat radiated will only be effective while it issues from luminous fuel, and luminosity depends upon quantity and quality of coal. A material saving might, however, be effected amongst the poor by the adaptation of the small Belgian stove, which travellers may have seen in use in Belgium. This is very serviceable; but the people of this country will long continue the old and wasteful grates, for they dislike stoves of all kinds.

A number of methods of economising fuel will suggest themselves to all large consumers of coal. No doubt many steamers will coal at foreign ports for a time, and similar expedients may be adopted in certain departments of manufacture. But if we sum up all of these savings, and look at them as hopefully as we may, the utmost early diminution which they may produce in our annual consumption will not, we fear, be very considerable, and the sudden loud cry for universal economy in coal will die into silence without an echo. Those who sagely counsel us to reduce our consumption and extraction within definite limits, such as eighty or fifty millions of tons annually, would do well to point out the methods and the probabilities of any such diminution. There is really little hope of it; a prohibition of exports would save most, and the most readily; domestic thrift would secure some considerable saving, but not so much as is vaguely expected; metallurgical and manufacturing economy is the most important, but not immediately the most promising, element of hope; so that we are reluctantly brought back to rest upon unwelcome conclusions. Even if Government were to buy up and work all our coal mines—which, however, it cannot and will not do—the difficulty would only be shifted from the

shoulders of coal owners and workers to the shoulders of
Government; for it is plain that the natural conditions
of the extraction and delivery of the mineral would
remain the same, while all the trouble of management,
and strikes, and social disorders would accompany the
transfer of the burden. The power of the State cannot
be brought to bear upon class contentions without changing
our system of Government.

Having thus adverted to some impracticable suggestions
for our present difficulties, let us take others and more
apparently feasible ones into brief consideration. What
are the present and prospective remedies for, or allevia-
tions of, the existing and future scarcity of coal? and
what can we do to prevent an aggravation of our present
calamity? Let us advert to the several probabilities
which we calculate upon from our present and past
experience.

In respect of human labour in coal pits, it is hard to
see how it can or will become continuously cheaper,
though it will perhaps materially abate its present
extravagant pretensions. We must exclude moral and
educational possibilities from the consideration; for so
long as Trades' Unions continue powerful, and working-
men remain inaccessible to argument and reason, it is vain
to offer them enlightenment. All, however, who have
opportunity should show these deluded miners that they
are working mischief not only to the nation but to themselves, and not merely moral but also pecuniary mischief.
If they reduce their amount of labour, by so much do
they shorten the supply and raise the cost of coal to the
public, and especially to their own order, the artisans and
the poor. By enhancing the cost of coal to consumers,
they will ultimately advance the cost of nearly every
necessary of life produced in this country, and living
will become dearer to colliers themselves. Their wages

will purchase less, and though these should be nominally higher, they may in time be relatively lower. This and other certain evil consequences, both personal and national, might be orally explained in lectures, or printed in simple tracts and largely distributed.

The pitmen being commonly an isolated class of workmen are strongly prejudiced, and pride themselves upon their labour being skilled and difficult of acquisition by navvies; and so it certainly is, more particularly in the thinner seams of coal, and in the deep mines, as well as in the inner recesses of deep mines. But the skill is chiefly amongst the hewers, and the other kinds of workmen are decidedly wrong in magnifying the skill required by them, for much of it is merely hard work. The art in hewing is that of bodily adaptation and posture, not that of delicate fingering and quick thought and eyesight. There are a dozen kinds of skilled labour greatly more difficult, demanding, indeed, much less physical exertion but far more of the superior qualities of head and ready-handedness. Well knowing what coal-pit labour is in all its forms, for we have seen it often in them all, we venture to affirm that a thoroughly good and patient colliery viewer could, under urgent necessity, drill and discipline some hundreds of willing agricultural labourers and workers, particularly if they were young and in all other respects suitable; and if the same viewer could keep the same men at the same work for a year or two, they might be made tolerably capable pitmen. If, therefore, the present colliers continue impracticable, and obstinate, and unreasonable, the coal owners must obtain the best available substitutes for them. Unquestionably a hewer who is educated to his work from boyhood is best; but as hewers of this kind do not largely increase while the demand for them does increase, employers must have recourse to another class. No doubt all who are drafted into pits from other classes of labourers might strike, and belong to Unions, and suffer and commit all sorts of mischief; but to enumerate such possibilities does not help us, nor does it hinder the desirableness in a time of necessity of trying this remedy. When our British colliery Othellos saw their occupation going or gone, they themselves would become more reasonable, and much more manageable for the future.

A far more agreeable compromise between the modern antagonism of labour and capital would be a union of interests. In this view the pamphlet of Mr. Briggs, named at the head of our article, is seasonable and instructive.   The Whitwood Collieries, situated near Normanton, and of which Mr. A. Briggs is the Managing Director, were worked by a private firm previously to 1865, and, says Mr. Briggs,

It is really almost impossible to overstate the virulence of the passions that were cited by the frequent disputes between masters and men.   During ten years we went through four strikes, lasting in the aggregate seventy-eight weeks; besides innumerable disputes and consequent interruptions to work.   During those ten years of almost daily recurring annoyances and anxiety, the firm went through the varied and painful experience of enforcing evictions from their cottages, guarding, by the aid of the police, non-unionists from the attacks of desperate unionists, the receipt of threatening letters on the model of those revealed before the Trades' Union Committee at Sheffield ; the whole culminating in a riot in the village on the night of September 24, 1863, and the consequent prosecution and conviction of some of the ringleaders at the following York Assizes. The pecuniary effect of these difficulties, as regards the employers, was that for several years barely 5 per cent. per annum could be realised on the capital embarked ; while most of the workmen were so impoverished, that in many cases they were compelled to dismantle their houses, and to sell property, the fruits of former labour, to obtain the means of subsistence during the continuance of the strikes. So disgusted were the owners of the collieries—not merely with the pecuniary sacrifice entailed, but also with the constant annoyance and anxiety of mind thus occasioned—that they had serious thoughts of disposing of them, trusting to find another investment for their capital, which might bring in a better return, and which would not, at any rate, lead to such incessant conflicts with the apparently inveterately adverse interests of labour.

Fortunately for them they did not sell their collieries, but converted the whole concern, in 1867, into a Joint-stock Company, with an encouragement in every form to the workpeople to become shareholders. Here was the co-operative principle proposed in a colliery company. Whenever the divisible profits accruing from the business should, after provision for redemption of capital, exceed ten per cent. on the capital embarked, all the employed, in whatever capacity, would receive one-half of such excess profit as a bonus to be distributed amongst them in proportion to, and as percentage upon, their respective earnings during the year in which such profits might accrue. Thus the workpeople became directly interested in the concern, and if they could aid in realising a larger profit they themselves would receive a share of it.

The effects of this change—the dissatisfaction of some, the conversion of others, the apathy and caution of many, and the final co-operation of all—are detailed by the Director. When the Company was enabled to declare a dividend of 12 per cent. on the capital for one year, and to devote a further 2 per cent. or 1,800*l.* to the formation of a workman's bonus fund, out of which was distributed an average bonus of 7½ per cent. upon their year's earnings among all the workmen who had properly qualified themselves, their eyes and hearts were opened, as well as their purses, and many of them had a 5*l.* note in their pockets for the first time, while some had two, the highest bonus paid to a miner being 10*l*. 18*s*. 10½*d*. upon his year's earnings of 109*l*. 8*s*. 9*d*. 'From that time,' adds the writer, 'we have gone on prosperously (up to 1870), dividing from 12½ to 13½ per cent. to our shareholders, and a proportionate bonus to labour.' In a very recent communication to ourselves in answer to enquiry, Mr. Briggs writes, 'Since that time the In- dustrial Partnership system has on the whole worked well. The production of the collieries is now at its maximum, and this in spite of advance in wages during the last two or three years of 40 to 50 per cent. We believe few private firms could say as much.' This is no doubt excellent when there is a profit to divide, but how would it be in bad years when the Company may have to bear a loss?

Nothing can be more satisfactory, and we commend this pamphlet (privately printed), and the very useful details which the writer has simply and clearly recorded, to the study of all coal owners and workers. The obstacles to a general introduction of the co-operative principle amongst pitmen are unhappily manifold; but looking at this particular instance, we find that one colliery concern, with about 100,000*l*. of share or invested capital, pays about 60,000*l*. per annum in wages, and that sum, therefore, falls to be considered as the average labour capital. Any further derivable profit, after the initial ten per cent. on the invested capital, would be appropriated to the payment of a future dividend or bonus on the whole 160,000*l*., being the aggregate of the united investment of monetary and labour capital. Thus, therefore, the proportion in which labour is to participate in profits will regulate itself, and the prosperity of the whole depends upon the recognition of the unity of interests, between the representatives in this undertaking, of capital and labour. Much remains to be considered, for which we refer to the pamphlet, only further observing the distinction, that this is not a mere co-operation of working men, but also a union of those who are commonly antagonists, and so far a successful merging of their antagonism.

We must either subordinate human labour in mining to capital, or we must harmonise it with capital, or supplant it by contrivance and mechanism. No other courses are open to us. We have long been trying to subordinate it, let us now try to harmonise it, or the third course becomes imperative. Many are now anxiously looking to this last course as increasingly imminent, and as perhaps ultimately imperative. Let us therefore offer a few remarks on this subject so far as they can be adapted to general intelligence.

At the outset, it is apparent that if mechanical coal-cutters could be extensively and successfully employed, several advantages would be gained besides supplanting the men. Not only would strikes be at an end, but ventilation improved, and the depth of mining probably increased, since most of the difficulties caused by the limit of human endurance under exhausting exertion would be at once overcome. Collateral benefits of several kinds, particularly the disuse of gunpowder to at least a large amount, would follow; and on this account the question of the possibility of their adoption becomes highly important. Can mechanism of any kind, whether partly or wholly employed in coal mines, or can improvements in applied mechanism, be made to supersede the labour of man ?

In thick seams like those of Staffordshire, in ten yard or even ten feet seams of coal, there is plainly no serious obstacle to the use of coal-cutting machines. You have only to cut and bring down, and deliver at the pit's mouth. But in thin seams and in the innermost passages

of long-worked pits the case is very different. Anyone
who has actually seen a hewer at work in the most
awkward places in thin seams, and no other person, will
at once understand the difficulties to be overcome. A
hewer in the worst 'faces' of coal-getting must squat and
twist and contort himself like a posture-master, and do
so for some hours at a time. This adaptation the human
body is capable of affording and enduring, and that most
wonderful vital machine, the human arm and hand, can
so twist and turn and harmonise itself to natural necessities as to show in this very circumstance its immeasur-
able superiority to any mechanism which the mind of man
has conceived. Watch, for instance, a northern hewer
making his 'jud' and his 'jenkin' in the coal seam, and
then turning and squatting, and sidling and squeezing,
and gasping and sweating, and picking and poking with
perfect bodily adaptation to mining exigencies, and you
will form such a conception of the peculiarities of the
work as will very much modify your expectations of a
mechanical substitute for the grumbling and perspiring
hewer.

Several coal-cutting machines have been brought under
the notice of colliery engineers and managers during late
years, and these have been considered and discussed at
their local meetings and by their committees. Some are
superior to others, but practical men soon perceive their
defects, which, indeed, are more or less inevitable. You
cannot convert the machine into the man, any more than
you can make the man a machine. In one respect it is
good that the machine has no free will. Then in another,
it is just the lack of free will which precludes free use.
So soon as we inspected several of these machines we
perceived their faults, while admitting their excellences.
Allowing for their defects, the question is, will they
win the coal? can they be made to do so by use, and

by a gradually acquired knowledge of needful improvements?

One or two of such machines have commended themselves more than others to coal engineers. That which at present appears to be one of the most promising is well worth a passing description. It is known as Gledhill's Patent Imperial, and has been successfully introduced by Messrs. William Baird & Co. in Scotland. The work is performed by an endless chain with attached cutters, driven round an arm which extends underneath the coal. When the machine is at work, it draws itself by means of the motive power of air, which is compressed at the pit's mouth to 35 or 40 lbs. per square inch, and is conveyed from the pit's mouth to the inner cast-iron pipes, and while at work it is attended only by three men. It is hopeless to render it quite intelligible without one or more drawings, and in fact even with these—and they lie before us as we write—it would be unintelligible to the unitiated in machinery and coal-mining. Yet it is easy to understand how such a machine having nine cutters of coal sharpened at each shift of work by removal daily to the surface for this purpose, and, when sharpened, again fixed in the chain, can proceed in its assigned duty. Mechanical ingenuity can certainly contrive to propel such cutters into a regular coal seam, to make those cutters of the best metal, to sharpen and readapt them, to fit them very nearly to the natural face of the coal, and to undercut with considerable efficiency. The colliery engineers wherever the machine is employed must of course see to the special adaptation of the machine to their particular pit.

There are, besides those collateral advantages over human labour by the employment of these machines to which we have already alluded, others of the same character which will be discovered in practice. One signal advantage is the saving of coal from waste. A man wastes coal in all operations and all directions. The hewer tears down the coal like a tiger, flings it into his heap, and at the same time into your mouth and eyes as a spectator of the hewing in the innermost recesses of a great pit filled with coal dust, and hence arise to the constant workers asthma and bronchitis. But the machine works without bluster or dust, without more of dust and without more of waste than is inevitable in all extraction of coal.

Two of these machines are to be tried at the famous Hetton Colliery near Newcastle, and we shall therefore in time hear of its success, or on the contrary. The Messrs. Baird are sanguine and quite assured, and tell us that the present work done by this invention is 300 to 350 feet, cut 2 feet 9 inches deep, in a shift of from eight to ten hours' work, and as the particular seam worked by them is 2 feet 10 inches thick, the yield is from 75 to 90 tons. As the cost of each machine is 200*l*., it would be easy to calculate the pecuniary results of their adoption. We do not anticipate that human labour in and about mines could be entirely dispensed with. Of the 300,000 colliers now employed a proportion would be retained. Those who propose these machines reckon that if universally employed in our coal pits, 60,000 colliers would suffice to raise our annual extraction of 120,000,000 of tons.

It would be well worth while to spend some thousands of pounds in inventing or perfecting coal-getting machines.[1] All interested persons will look anxiously to the results of the intended Hetton Colliery adoption of

[1] We refrain from noticing several already proposed coal-cutting machines, because they are subject to mining and mechanical judgment. A few years of trial will decide the question, and it will be one of the few benefits traceable to our present coal deficiency, if the old proverb shall be realised in this matter, and ' Necessity prove the mother of invention.' For a like reason we merely name the advantage of largely adopting coal-wash-

them. If they succeed in the largest pits producing the best northern coal, the end of our distressing coal famine is not far off, and, what is best of all, it cannot suddenly and unexpectedly return. Let us remember that although machinery will not prevent, but rather by its ingenuity accelerate, bituminous and carbonaceous bankruptcy, it will on this very account enable us to calculate and forecast the *ultima dies* and to prepare for its arrival.

If all mechanical substitutes fail in coal-getting, and if human labour becomes and continues still more and more intractable, there will remain only two other principal or partial remedies : one is the substitution of another fuel for coal, the other an importation of coal from other countries.

No prospect of a natural substitute for coal dawns upon us at present. That sun whose beams shone upon the primeval vegetation and originated coal shines upon all alike—the good and bad, the near and the remote ; but the coal it helped to form is only locally stored. If we turn to Peat, there are immense and readily accessible deposits of that useful substance, and peat-compressing machines are being perfected and vaunted. But it will never, for other reasons besides cost, extensively supply the place of coal. Nature gives us nothing like good coal ; science holds out no hope of anything like it ; and the combinations and decompositions of chemistry are at present in this direction only vaporous and vague.

It is melancholy to contemplate a necessary importation of coal to a country which by its possession and utilisation has dominated the manufacturing world. It will be indeed a kind of moral retribution when we, the great and

ing machinery, by means of which many thousands of tons of small and refuse coal, long lying useless in coaling vicinities, may be cleansed from earthy matter, shale, and pyrites, and reduced and screened, and made marketable.

prodigal exporters of coal, or rather our less fortunate descendants, shall come to beg abroad for the mineral we have for a century been sending away. We have had coal enough and to spare, but we have squandered our inheritance, and alas! may be severely punished.

Speculating upon such a dismal futurity, let us in imagination unroll a coal-map of the known world before us. Such a map could never be stereotyped, because geographical exploration and topographical surveys reveal to us more coal than we even a few years since knew to be on our earth. Moreover a map of the surface would be of little use to us; we require particulars of thickness, depth, quality, and deposition.

Very probably we shall commence importing foreign coals by having recourse to the Belgian coal field, which lies near to us, although at present the Belgian demand presses on the supply. This forms a part of a long and comparatively narrow series of basins, extending about seventy-five miles from east to west, and lying in about equal proportions within France and Belgium, the latter country possessing a field of about forty miles in longitudinal extent, and eight miles wide in the mean, with an area of 326 square miles of productive coal measures. Most of the seams, however, are thin and less than two feet thick on an average. About 120 seams are now well developed, but the production is limited, in comparing the number of working people with what a similar number would produce in a British mine. There may be fifty thousand persons altogether employed in the Belgian collieries, of whom about forty thousand are engaged underground and ten thousand above ground. The total production in 1864 was reported at 10,000,000 tons of coal, whereas in 1850 the total extraction did not amount to quite six millions. It is now perhaps twelve millions.

We may also resort to the coal fields of Westphalia, which ought to be generally known, as they contain more than sixty beds of workable coal in seams of from three to nine feet thick, comprising a total thickness of about 200 feet of pure coal of various kinds, and extending over a surface of about 200 square miles. It is estimated that there are about forty thousand millions of tons of coal in this entire field, and calculations would probably show that this coal might even now be imported by sea to the port of London at a much less price than we have been recently paying for our fuel. In 1867 the Ruhr Coal Basin produced ten and a half millions of tons of coal, while in 1851 it did not yield two millions of tons. Railways and low charges have promoted its development.[1]

In the future, probably our country will obtain coal from the British North American Provinces, which contain about 8,000 square miles of the mineral. Remaining under our own dominion, this mineral resource will be at our immediate service; and the coal in Nova Scotia is already attracting considerable attention. Were it not for the impediments of distance and carriage we might at once avail ourselves of these deposits, some of which are now extensively mined, but cannot be wrought at a price to meet our home necessities. The element of cost is the forbidding element, even in our own remote dominions; and, in fact, a mountain of coal at three or four guineas a ton would be of no more advantage to us than Mont Blanc with thousands of tons of useless snow.

[1] We can only name the Saarbrück coal field in Rhenish Bavaria, although it has an area of nearly a thousand square miles. Dr. H. von Dechen has published a detailed account of the coal fields in the Rhine Provinces and Westphalia. The development of its coal mines will probably prove of prime importance and interest to Germany in the future, and perhaps also to us.

For a like reason the vast coal deposits of India are of small present value. They are full of hope for the future progress of India, but a cold comfort to us. Dr. Oldham and his coadjutors have done good work in estimating that about sixteen thousand millions of tons of coal exist in all the known Indian coal fields. The extension of the railway system of India will of course be largely aided by the discovery of native beds of coal; and we shall be relieved from the necessity of exporting British coal for Indian locomotives and navigation. But that is the extent of the benefit to ourselves. Few persons who speak or write upon the coal famine which at present perplexes the whole nation, appear to remember that the existence of immense remote deposits of coal is of little or no value to us in our existing emergency. Looking at the long future these are sufficiently important, but looking at home they offer no solution of our difficulties, and no hope of low-priced coal next winter. For the remote future alone do they afford hope. In the last number of this Review (p. 243), we gave a brief notice of the discovery of the extensive coal fields in East Berar, comprising an estimated aggregate of 480,000,000 tons of coal. The subject of Indian Coal Resources would of itself demand a separate article.

It has only of late years been made known that the coal fields of China extend over an area of 400,000 square miles; and a good geologist, Baron Von Richthofen, has reported, that he himself has found a coal field in the province of Hunan covering an area of 21,700 square miles, which is nearly double of our British coal area of 12,000 square miles. In the province of Shansi, the Baron discovered nearly 30,000 square miles of coal, with unrivalled facilities for mining. But all these vast coal fields, capable of supplying the whole world for some thousands of years to come, are lying unworked;

and it appears from some late observations in the House of Commons, that certain overtures lately made for bringing these valuable deposits into mining development were rejected. The Chinese authorities and people will prefer to pay for the coal imported rather than use their own. They are ignorant of good mining, object to it politically, and suffer all the natural difficulties which other nations have overcome, to 'frighten them into inveterate obstinacy and industrial inactivity.

The greatest resource for coal in the future will be the deposits of the United States of America, which have an area of coal formations extending over 500,000 square miles, under which the productive or workable area has been calculated at 200,000 square miles. We can count up more than twenty coal fields in America, some of which are small and others very large. Pennsylvania possesses no less than 12,656 square miles of bituminous coal and 470 square miles of anthracite; while West Virginia has 15,000 square miles; Illinois, 30,000 square miles; Michigan, 13,000; Iowa, 24,000, and Missouri 21,000 square miles of coal. Add to these the great coal fields lying within the Ancient Appalachian Basin, amounting in all to 203,000 square miles.

It must not be thought that the small extent of the areas of anthracite in America is a measure of their value, for their value is in inverse proportion to their area. In many respects Pennsylvanian anthracite is peculiarly serviceable; it is more dense and compact than other kinds, and a pure specimen will yield from ninety to ninety-five per cent. of carbon. On the north side of our South Welsh coal fields we possess an anthracite resembling that of Pennsylvania, and which is therefore valuable as steam coal. At this present time the Americans would perhaps value their 470 square miles of anthracite in Pennsylvania above their 200,000 square miles of bituminous coal lying elsewhere, for the beds of anthracite are situated in the vicinity of numerous great and prosperous cities, and have done for all that region something like what our coal fields have done for us. There are now 12,000,000 of people deriving their chief supplies of coal from these deposits, and the same process is going on there as here. The present 12,000,000 will increase, and the present produce of anthracite likewise, and they will increase in accelerated ratios. Taking the present annual anthracite production at about ten or twelve million tons, this will soon advance with the enlarging demand, and we can readily anticipate both the simultaneous growth of population and anthracite mining. A dense population rapidly extending over an area twice as large as that of Great Britain, will eventually bring up the extraction of anthracite to a high amount, and the cost will increase until the people begin to work their bituminous coal, which will always preserve them from a coal famine like our own, and at the same time tend to bring more and more of their adjoining deposits into mining and to market.

Numerous momentous results depend upon the successful solution of the problems adverted to in this article. That our commercial prosperity is founded upon the possession and use of coal is sufficiently known as a general truth, but the whole state of the case is not so widely understood. Great Britain has become the working coal field of the world. We have for some years been raising more than half the total coal raised in all parts of the globe, and we have recently raised considerably more than half that total, which in 1866 was estimated at 170,430,544 millions of tons. If the whole world may now be supposed to raise annually about 200,000,000 of tons, we are raising about 120,000,000 out of the 200,000,000. Yet with this immense extraction we have only about 32,000,000 of inhabitants in the United Kingdom.

Let us contrast our case with that of the United States.

That vast region has a present population of not much more than 39,000,000, for the census of 1870 returned it as 38,558,371. Last year (according to a recent advice) the total production of coal in the States was 41,491,135 tons. In 1865 the total production in the United States was 11,324,207 tons, and of anthracite 11,532,732 tons; altogether making 22,856,939 tons; and the capital then employed in coal-mining throughout the United States was 40,000,000 of dollars, while the capital invested in railroads and canals penetrating those coal fields, made principally for their development, and sustained by the coal trade, amounted to 170,000,000 of dollars. This large increase is a striking example of prosperity; but the Americans have one serious drawback, namely dear labour, and they cannot as yet compete with our cheaper labour. Should, however, our mining labour continue at any such cost as recently, the wages in America may become actually less than ours or equal to them. Then the present relative conditions may be reversed; *we* have possessed far less coal, and enjoyed far better and cheaper means of extracting it; *they* have had the most coal and have only required as cheap or cheaper labour than ours, and equal skill in mining, to turn the balance in their favour.[1] The Superintendent of the American

---

[1] The question is too large for present discussion, or we should show grounds for believing that the balance is now turning against us, and that just as coal and iron have become dearer here, the Americans have resorted to their own iron and wrought their own coal. In 1871 and 1872 they exhibited great energy in both departments, and as a consequence have bought less of our iron. In the first two months of 1872 they received 88,430 tons of our iron rails, while in the first two months of this year they took only 48,901 tons; thus showing a decrease of 39,529 tons; and as American orders largely go to Wales, we at once arrive at an important conclusion on

Census thinks that the population of the United States in the year 1900 will number 100,000,000. If by that period the balance should be turned in their favour, then the vast demands for coal will have stimulated a corresponding extension of coal-mining; and the Americans must open coal fields of far wider extent than our own.

The course of manufacturing supremacy, of wealth, and of power is directed by Coal. That wonderful mineral, of the possession of which Englishmen have hitherto thought so little, but wasted so much, is the modern realisation of the philosopher's stone. This chemical result of primeval vegetation has been the means, by its abundance, of raising this country to an unprecedented height of prosperity, and its deficiency might have the effect of lowering it to slow decline, while by greater abundance it raises another country in the Far West to a prosperity possibly greater than our own. It supplies food, force, heat, light and motion—wonderful alike in its geological origin and in commercial and national influences. It raises up one people and casts down another; it makes railways on land and paths on the seas. It founds cities, it rules nations, it changes the course of empires. And along with all this physical and social efficiency, it reads us a grave and solemn admonition. There is a moral purpose and retribution in all its vicissitudes. Prodigality, wastefulness, lack of prudent calculation, social selfishness, embittered class interests, and the national neglect of social and moral as well as physical laws in relation to this one indispensable gift of nature, will assuredly bring retributive justice upon us all or upon our posterity.

the suicidal character of strikes in that country. It has recently come to light that in the English market iron is 3*l.* per ton dearer than in the American market.

II.

ON THE COAL FIELDS OF NORTH AMERICA AND GREAT BRITAIN.[1]

WE had occasion to refer in a recent Number, for another purpose, to the magnificent and elaborate work which Mr. Rogers, under the liberal patronage of the state of Pennsylvania, has recently given to the world. Our present object concerns exclusively the Essays on the Coal formations of various countries, which are by no means the least interesting and important portion of these volumes. The history of Coal affects, in the highest degree, the entire social condition of our species; and we propose to consider it in the following pages, not so much with reference to its geological characteristics, as in connection with the prodigious services which this mineral renders to civilisation. In those seams of combustible matter, which the industry and ingenuity of man have discovered and worked in various parts of the globe, lies the latent force which gives life to the steam-engine. Heat, motion, power, and that wonderful energy which propels in a thousand forms the mechanism of modern society, are all concentrated here; and the geological revolutions which reduced the primeval forests and morasses of the globe to this condition, were preparing, in the incalculable distances of past ages, that new element

---

[1] *Essays on the Coal Formation and its Fossils, and a Description of the Coal Fields of North America and Great Britain, annexed to the Government Survey of the Geology of Pennsylvania.* By Henry Darwin Rogers, State Geologist. Edinburgh and Philadelphia: 1858. 3 vols. 4to. with Plans.

which was one day to make man the master of earth, of water, and of fire. To our mind there is nothing more indicative of the eternal forethought which framed the structure of the world, than the fact, that perishable organisations, which flourished thousands of years before the existence of man, should have become subservient to the latest applications of human skill. The power of calculation can hardly measure the stupendous addition made by this force to the dynamic power of man; but we may borrow the estimate given by Professor Rogers, of the value of the coal fields of this small island—a very small portion, as we shall presently see, of the vast coal fields which stretch across the globe.

Each acre of a coal seam, four feet in thickness, and yielding one yard net of pure coal, is equivalent to about 5000 tons, and possesses, therefore, a reserve of mechanical strength in its fuel equal to the life-labour of more than 1600 men. Each square mile of one such single coal-bed contains 3,000,000 tons of fuel, equivalent to 1,000,000 of men labouring through twenty years of their ripe strength. Assuming, for calculation, that 10,000,000 of tons, out of the annual produce of British coal mines, are applied to the production of mechanical power, then England annually summons to her aid the equivalent of 3,300,000 fresh men pledged to exert their fullest strength through twenty years. Reducing this to one year, we find that England's actual annual expenditure of power, generated by coal, is represented by that of 66,000,000 of able-bodied labourers. This is a representation of what really exists in another form; but if we proceed so far as to convert the entire latent strength resident in the whole annual produce of our coal mines into its equivalent in human labour, then, by the same process of calculation, we shall find it to be more than the labour of 400,000,000 of strong men, or more than double the number of adult males now upon the globe!

An element in the above calculation is one of the most humiliating comparisons that can be drawn between human and mechanical power. If we estimate a lifetime of hard human work at twenty years, giving to each year 300 working days, then we have for a man's total dynamic efforts 6000 days. In coal this is represented by three tons; so that a man may stand at his own door while an ordinary quantity of coals is being delivered, and say to himself: ' There, in that waggon, lies the mineral representative of my whole working life's strength ! '

In such aspects as these, how momentous to ourselves is the natural possession of coal—of the fuel ever ready at a moment's preparation to generate a power the very opposite of its own nature—a power that transcends all others yet known to be applicable to mechanical movements ; that disdains narrow imprisonment, and wings us or wafts us over land and sea—that daily draws up from the deepest pits more and more of the mineral fuel that gave it birth and impulse—that makes tens of thousands of wheels and spindles to revolve incessantly—that causes raw materials to be wrought into airiest fabrics or solidest structures—that transports navies and armies, changes the character of warfare by accelerating the transfer of men and the munitions of war, decides the fate of battles, and determines the destiny of nations. How momentous, we repeat, is the possession of the generator of all these movements ! In our extensive beds of coal we have, in fact, the motive power of the world, stored up for us in the most compact and suitable form. In coal we have the world-moving lever of Archimedes, and it may be said that the steam engine is the fulcrum. Few economical enquiries, therefore, can assume greater importance and higher interest than those which are connected with our coal deposits. Under what geological conditions do we find them in our own and other countries? what

approximation can we form to the amounts possessed by
the principal countries of Europe and America? More
particularly, what are our British coal possessions, at what
rate have we mined, and are we now mining them, in
what directions and for what purposes are our supplies
distributed, and can we arrive at any approximate
estimate of the vast quantities we distribute to domestic
use, to manufactories, and to foreign countries? Finally,
directing our attention to the qualities of certain coals,
what are the conditions of a good steam coal; to what
extent are steam coals possessed by ourselves and other
nations; and how far does the possession of such qualities
of coal affect the conditions of naval warfare and the
prospects of victory?

The great Carboniferous series of rocks which derives
its name from the coal found in them, is both geologically
and economically important, and we may add, as regards
our own country, artistically interesting; some of the
sweetest river and valley scenes in our land are indebted
to the character of these rocks for their scenic beauty and
even grandeur. They occupy an immense tract in
Northumberland, Durham, Yorkshire, and South Wales;
and in Ireland, the greater part of its plains, and of the
broadly undulated interior, consist of the mountain limestone, covered in some places by the coal measures, and
in others supported by the old red sandstone. Although
mineralogically the same rock is termed Carboniferous or
coal-bearing, yet this is chiefly because it is the principal
rock associated with the series of coal-bearing beds, or as
they are called, the *coal measures.* The Carboniferous
limestone itself, in south-western England and South
Wales, includes no coal of consequence or amount; and
except in some rarer and higher parts of this limestone
formation, not even small coal seams can be traced in it
n the districts named. It contains, however, a rich
assemblage of organic remains which are essentially of

marine character, and the whole is evidently a marine formation. The same general conditions appear to have prevailed in the British Islands, as far north as Derbyshire and North Wales; but as we advance further northward, the coal beds become more intermingled with the mass of supporting calcareous deposits, from which we may infer that in the northern portions of this area, the natural conditions favourable to the growth and entombment of the vegetation from which coal was produced, commenced at an earlier geological period than in the southern parts of the area indicated.

The coal measures themselves, or the series of beds intimately associated with the seams of coal, consist of a number of strata of alternating sandstones, shales, and coals. But the pure beds of coal in this great aggregate are insignificant in comparison, for in all they only amount to a thickness of 47 feet 9 inches, and are, therefore, merely in the proportion of 15 of coal to 38 of rockmeasures. Later researches give an aggregate of 76 feet of coal. Now as none of the seams of coal separately exceed 6 feet of available thickness, while many of them are not more than as many inches, the stranger who for the first time sees a Newcastle coal seam underground, can scarcely bring himself to apprehend that the vast expenditure he beholds, and the array of machinery and labour in action, are all directed to the excavation of a thickness of a few feet of coal, and least of all, when the seam is barely 6 inches thick. In tracing this great series of rocks through our own country, we, in fact, trace the coal; for the pure bituminous coal has never with us been largely found apart from that series; that is its true geological position. The exceptions in the oolitic coal seams of Brora in Sutherlandshire, and at Gristhorpe in Yorkshire are of trifling importance. Had this now familiar fact been previously recognised, many thousands of pounds vainly expended in the search for

coal, would have been saved, and many ridiculous enquiries as to the possibility of finding it in particular
places would have been spared.

The original formation of coal is one of the most interesting enquiries in theoretical geology. That coal was
composed of primeval vegetation is demonstrated by the
experiments of such chemists as Liebig as well as by the
evidences it carries in its own existence, and in its neighbouring sandstone. A coal bed is in fact a *hortus siccus*
of Old-World vegetation,—and this is the more strikingly
seen when some diligent collector, like Mr. Hutton in
the north, gathers together in the course of years a
cabinet of coal plants. To inspect such a collection is to
behold compendious evidence of the existence of a most
ancient and luxuriant flora, from the remains of which we
are at this day deriving heat, light, and power. Careful
research into the character of the coal plants enables us
to restore in imagination the standing and growing plants
of that remote era. We may depict huge trees of strange
forms ; thick hedges of tall reeds, with glossy stems and
radiating pointed leaves ; gigantic club-mosses and innumerable ferns, overtopped by trees like pines; and a
strange plant resembling an immense coach-wheel within
its rim, so that the boughs shoot out horizontally on all
sides like spokes from the nave,—the central portion of a
Stigmaria ficoides, once floating near what is now Newcastle ; and in our own Scottish coal field of Fife a fossil
trunk, supposed to have belonged to a gigantic Araucaria,
has been extracted from the earth at no great distance
from the richly wooded hills of Raith.

The more we can discover concerning the fossil plants
in and near the coal beds, the nearer shall we approximate
to a solution of the problems connected with the formation of the coal itself. Much knowledge on this subject
has been gradually accumulated. Erect fossil trees have

been found in some parts of Europe as well as in our own
country.   Five examples of fossil roots and trunks were
discovered erect, as they grew, on the Bolton and Manchester Railway, and above them was a seam of coal two
feet in thickness.   In the extensive coal field of South
Wales, Mr. Logan affirms that there is no instance of a
seam of coal occurring without a bed of underclay which
abounds in remains of the marshy plant *Stigmaria ficoides*
(now considered to be the roots of a *Sigillaria*, or probably other plants), and he conceives, from its abundance,
that it must have been the chief component of the bed of
coal.   He found the same kind of underclay and the
same plants in the Pennsylvanian coal fields.   Moreover
vertical stems of plants are found at more than one geological level, and in coal districts in Nova Scotia and
Cape Breton, in several of the planes of vegetation stems
are still seen standing in their places of growth above
each other, to the extent, it is supposed, of fifty or even
one hundred ancient forests buried one above another.

A glance at the beautiful plates of Professor Rogers'
work recalls to us very nearly the same fossils from the
coal formation of America, as may be seen in the plates
of the plants *Fossil Flora* of Lindley and Hutton.   We
could find many or even most of the American fossils,
with slight specific differences, paralleled in our British
cabinets.   The variations in the vegetation are evidently
only such as might be found at great distances under
similar conditions of climate.   Two hundred and twenty
species of plants from the coal measures and carboniferous
rocks of Pennsylvania, yielded more than one hundred
species which were entirely identical with species already
recognised in the European coal fields.   Fifty others
showed differences so slight that a fuller comparison with
better specimens may result in their identification ; while
even the new species, which seem to be restricted to that

foreign field, bear in every instance a close relationship
to European forms.   It is, therefore, upon the best
evidence that we pronounce a remarkably near affinity to
exist between the carboniferous fossil flora of North
America and of Europe.   This is but another confirmation of the closely analogous conditions of the ancient
coal era; and we might even show how very close the
affinity is in other particulars, as that the most common
species of plants in the coal flora of Europe are probably so in America, while scarcity in the one continent
corresponds to scarcity in the other.   It is also confirmatory of this view that the fossil flora of the oolites
of Scotland, to use the words of Hugh Miller, ' in its aspect
as a whole greatly resembles the oolite flora of Virginia,
though separated in space from the locality in which
the latter occurs by a distance of nearly four thousand
miles.   There are several species of plants common to
both ; both too manifest the great abundance in which
they were developed of old by the beds of coal into which
their remains have been converted.'

The abundance and vast size of certain coal plants
have formerly been thought to imply an extremely hot as
well as a moist and equable climate ; but further investi-
gations of the structure and relations of these plants
rather indicate a very different climate from the present
in that period than a very hot one.   The prevalence of
ferns points to an extremely humid, and at the same time
an equable and even temperate climate, without severe
cold.   Such is the view of many eminent geologists at the
present day, and it is derived from observation of the
conditions now favourable to similar tribes of plants, as
for instance in the islands of tropical oceans, and of the
southern temperate zone.   An opposite conclusion, at
least as far as regards tropical climate, has been arrived at
by M. Lesquereux, who assisted Professor Rogers, and
also by Professor Brook.   That gentleman affirms that

' nothing can authorise us to admit these atmospheric influences as very different from what they are now.' With
reference to the immense trunks of trees, perhaps of fern
trees, to which we, now find an affinity only in the tropical
regions, it may be said, how is it possible to account for
their growth in our latitude if we do not admit of a great
change of temperature? M. Lesquereux replies that in
the peat bogs of northern countries, of Denmark and
Sweden, much larger trunks of trees may be found than
those which have been discovered in ours; but the true
fern trees (*Caulopteris*) are very scarce in the coal; and
that as to presumed difference of temperature, the clearing
of valleys and the drainage of lands may cause a climate
not really to become colder, generally speaking, though
the extremities of temperature are more distant; that is,
it is colder in winter and hotter in summer. The degree
of this difference regulates the vegetation of a country;
and it is sufficient to afford us the reason of the difference
of the type of vegetation between the coal period and our
own, if we ' admit that the continents were less extended,
and only low islands entirely covered with marshes.'
Further, it is thought that all the physical phenomena of
our time were then in activity. There are in parts of the
American beds evident marks of drops of rain and of
hail, and also cracks caused by dryness under a burning
sun. In layers of coal, the thickness of which scarcely
exceeds the twelfth part of an inch, there are the proofs
of an annual decay, and of annual heaping of the plants
exactly as may now be traced in peat-bogs. In the bituminous coal of Ohio the annual growth of the coal is well
marked by the thin layers, which are about one-twelfth of
an inch in thickness.

If this could be accepted as a chronological unit, we
should be enabled to arrive at some conception of the lapse
of time demanded for the formation of seams of coal, and
of the vast abundance, as well as bulk, of the vegetation

of the carboniferous era. Not only were there arborescent
plants which attained a height of sixty feet, but the im-
mensity of the mass must have equalled, if not surpassed,
the luxuriance of growth. What an amount of vegetable
matter must have been required for the total seams of any
of our coal fields, and particularly for the Staffordshire
' thick coal,' which attains a thickness of thirty, and sometimes nearly forty feet ! Calculations have been made by
Mr. Maclean respecting the quantity of woody matter
which may be supposed to have entered into the composition of a given bulk of coal; the result of which is that
one acre of coal, three feet thick, is equal to the produce
of 1940 acres of growing forest; and that if the wood all
grew on the spot where its remains now exist as fuel, the
coal bed of the above dimensions would be the consequence
of a forest growth of 1940 years. Even if we grant a
rapidity of vegetation like that of a tropical climate, still
we should demand at least 1000 years for the formation of
one such coal-seam ; and for the thirty-six yards of coal
in the Mid-Lothian coal field, a period of at least 36,000
years. What time, then, would be requisite for the aggregation of such a coal deposit as that at Saarbrück,
where 120 beds are superposed on one another, exclusive
of many which are less than one foot thick; and what
time untold must we allow for the formation of the vast
deposits of America and Nova Scotia ? As Hugh Miller
observes : ' All the forests of America gathered into one
mass would fail to furnish the materials of a single coal-
seam equal to that of Pittsburg.' And Sir C. Lyell
remarks, with reference to the great beds of anthracite
coal, between forty and fifty feet thick, quarried at
Mauch Chunk (the Bear Mountain), in Pennsylvania:
' The accumulation of vegetable matter now constituting
this vast mass may, perhaps, before it was condensed by
pressure, and the discharge of its hydrogen and oxygen,

and other volatile ingredients, have been between 200 and 300 feet thick.' Confining ourselves even to the estimated total quantity of coal-contained in the entire known deposits of our own country, in what words or figures can we represent to ourselves the vegetation requisite to compose our 5,400 square miles of coal area, which is calculated to contain 190,000,000,000 tons of coal. Our minds are baffled in aiming to comprehend the bulk of original material, the seasons of successive growth, and the innumerable years or ages which passed while decay, and maceration, and chemical changes, prepared the fallen vegetation for fuel. In descending the shaft of a coal pit we shoot down in five minutes through a succession of beds, which represent a duration immeasurably surpassing the whole period of man's existence upon this globe. Mark off man's terrestrial duration upon a vertical scale composed of a dozen seams of coal, and the whole human age would form but an unit, though you should divide the scale minutely. Yet it is for this creature of a day that the primeval forests grew, the mighty ferns waved their fronds, the marshy plants spread their succulent leaves and stems, the favouring sun shone, the heavy rains descended, the hurricanes uprooted trees, solid growth succumbed to slow decay, and all the secret but sure resources of the laboratory of Nature were brought into activity to reduce the fallen or crippled vegetation into a carbonaceous and bituminous condition, and to prevent its admixture with arenaceous and deteriorating ingredients, to run it out into long and level or gently curved deposits, to pack it into solid sandstone cases and under huge shady covers, and to store it in the smallest compass by the mighty pressure of ponderous rock-presses!

We have before us, then, the ascertained facts that immense accumulations of vegetation were collected; that

this vegetation consisted largely of *Sigillariæ*, the root-
lets of which are met with in great abundance in the clays
representing the floors on which the coal seams rest; also
of reed-like plants, such as calamites and lepidodendra;
that coniferous trees (those of the fir tribe), do not form a
considerable element in the composition of coal; and that
the conditions under which the coal plants flourished
appear to have belonged to such swamps as were covered
to a great extent with water, rather than to anything in
the form of peat bogs.

As to the particular mode of aggregation, we may find
modern and existing illustrations in the timber which is
drifted down by great rivers, and is often arrested by
lakes.   This may sink after being water-logged, and may
become imbedded in lacustrine strata, if any be forming
in the locality.   A portion, however, will float on and
reach the sea.   We have an example of a vast accumulation of vegetable matter now in progress under both these
conditions in the course of the Mackenzie River, in
America, as noted by Sir C. Lyell.   Again, in an arm
of the Mississippi, drift trees, collected in thirty-eight
years, formed a continuous raft ten miles in length, and
eight feet deep.

Dr. Richardson[1] describes a still more illustrative
instance in the enormous annual amount of drift timber
brought down in the Slave Lake, which vies in dimensions
with some Canadian lakes.   There the trees retain their
roots, and being loaded with earth and stones, readily
sink, especially when water-soaked.   They there accumulate in eddies, and form shoals, which ultimately enlarge
into islands.   A thicket of willows covers the new-found
island as soon as it appears above water, and their fibrous

---

[1] Geognostical Observations on Franklin's Polar Expedition : cited by
Sir C. Lyell in his 'Principles of Geology.'

roots serve to bind the whole family together. Sections of these islands are annually made by the river, assisted by the frost; and it is interesting to study the appearance of the trees, according to their different ages. The trunks gradually decay, until they are converted into a blackish brown substance resembling peat, but which still retains more or less of the fibrous structure of the wood. Layers of this substance often alternate with layers of clay and sand, the whole being penetrated to the depth of four or five yards by the long fibrous roots of the willows. A deposition of this kind, made with the addition of an infiltration of bituminous matter, produces an excellent imitation of coal, with impressions of the willow roots. One of the most interesting resemblances to our coal measures observed in this place was the horizontal slaty structures presented by the old alluvial banks, and the *regular curve* which the strata assumed from unequal subsidence. The Slave Lake itself must, in process of time, be filled up by the matter conveyed into it daily from the Slave River.

Other situations are known where vast vegetable accumulations occur. In Iceland an immense quantity of birch, trunks of pines, fir, and other trees, are thrown upon the northern coast of the island, especially upon north Cape or Cape Langaness, and are carried by the waves along these two promontories to other parts of the coast, so as to afford sufficient wood for fuel and for constructing boats. The bays of Spitzbergen are in like manner filled with driftwood, which accumulates also upon those parts of the coast of Siberia which are exposed to the east. This wood consists of larch trees, pines, Siberian cedars, firs, and other woods; and the trunks appear to have been swept away by the great rivers of Asia and America. Scientific travellers are adding other observed phenomena of the same kind to our store of illustrations.

The mode and vastness of accumulation being in some measure explained, the next problem is the original formation or transportation of the immense bulk of vegetation thus gathered together. The two leading opinions on these points have been what are termed the 'drift' and the 'peat-bog' theories, and fierce geological battle has been done for both of them during the last thirty or forty years. According to the peat-bog theory, forests and jungles grew in the present coal localities, decayed into peat mosses, suffered subsidence with the land, which thus became the basin of a lake or estuary into which broad rivers conveyed mud and sand, out of which were gradually consolidated the now overlying and underlying shales and sandstones : and, during this period, the vegetable material bitumenised and mineralised into coal. The same area was again raised from the waters, was the ground for luxuriant vegetation, was again submerged, and again covered with successive depositions of shale and sandstone. An alternating subsidence and elevation are, in this view, presumed to have taken place for every seam of coal we find ; so that the coal seams in any of our deposits may be regarded as so many land wastes, and the sandstones and shales as river wastes, and in both of these we now possess a kind of amphibious chronology, a well-marked scale on which we may now read off the successive periods of sunshine and flourishing vegetation, of decay and desolation, of disappearance and submergence, of re-appearing and returning luxuriance, and of another era of upshooting tree-ferns, outspreading club-mosses, and broad flag-like foliation.

The drift theory, however, does not permit us to indulge such visions to the full extent. It admits of partial and limited submersions and elevations of land, such indeed as are now taking place on our earth, and also of dense jungles and peat mosses experiencing the same submersions; but its distinctive features are that the main bulk of the coal measures were deposited as drift and silt on lakes and estuaries, that the chief constituent vegetation was imported by rivers and inundations into such estuaries, and that numerous rivers might discharge their several freights of plant-remains or of mud and sand, into one estuary. These transporting rivers were themselves exposed to periodical inundations like the Nile and the Ganges; during the intervals between which the vegetation rank and rapid grew and closed up the deltas, and then furnished an important addition to the inland drift, the whole finally travelling down to form coal in the lakes or estuaries.

It has been felt that the objections to either theory are formidable. For the one theory the submersions and elevations supposed are too numerous, and the layers or 'partings' of sandstone or shale in some beds of coal, together with fossil shells and fishes, are too apparent and frequent to allow of only super-aqueous growth. Against the other theory there are, perhaps, still stronger objections, such as the evenness, general regularity, and frequently unmixed purity of coal seams. A combination of both theories, and an admission of the occasional prevalence of the phenomena peculiar to each, has been entertained as a solution of many difficulties; but there still remain some which perhaps our present knowledge does not suffice to remove. Ingenious but now perhaps untenable speculations have been offered to account for such remarkable facts as the uniformity of external condition over the extensive areas now occupied by coal fields in both hemispheres; the similarity of vegetation at that era over wide regions now temperate, tropical, and arctic; the prevalence, in short, of not a few conditions which now appear to be almost incompatible with known existing agencies and ascertained causes.

To account for some observed facts which the theories just adverted to do not exactly meet, Professor Rogers puts forth a modified theory of his own, which is in substance as follows :—The period of the coal measures was characterised by a *general* slow subsidence of those coasts on which the vegetation flourished. This vertical depression was, however, interrupted by pauses and gradual upward movements of less duration and frequency. These nearly statical conditions of the land alternated with great paroxysmal displacements of the level. During periods of rest or gentle depression, the low coast was fringed by great marshy tracts or peat-bogs, derived from and supporting a luxuriant growth of *Stigmaria*, while along the landward margin, and in drier parts of these sea-morasses, tree-ferns, conifers, and other arborescent plants grew profusely. In this condition, constant decomposition and growth of the meadows of *Stigmaria* produced a uniform and extended stratum of pulpy peat. To this scattered trees contributed occasional deposits of leaves and fronds and fallen portions which lodged in the marshes, passed to the pulpy state and ultimately formed coal, or preserved in several instances their vegetable structure. This view of gradual accumulation from *Stigmaria* and from deciduous contributions accounts for the marked infrequency and yet occasional occurrence of fossil *trunks* standing upon or in the coal, an instance of which in Lancashire we have previously mentioned.

Now suppose an earthquake, with undulatory movements of the crust of the earth, disturbs the level of the wide peat morasses, and adjoining flat tracts of forest on one side and the shallow sea on the other. The ocean, as in all earthquakes, draws off its waters for a brief time from the great Stigmariæ marsh, and from all the swampy forests which skirt it, and by its recession stirs up the muddy soil, drifts away the fronds, twigs, and smaller plants, and spreads them together with the mud over the surface of the bog. Hence come our laminated shales composing the immediate covering of the coal seams. Presently the sea rolls in with impetuous waves and prostrates everything on land. The entire forest is uprooted and borne off upon its foaming surges. Spreading inland, it washes up the soil, abrades fragmentary materials, takes them up and rushes out again with irresistible violence towards its deeper bed, strewing the products of the land in a coarse promiscuous stratum. It alternately swells and retires with surpassing energy, it oscillates tempestuously, spreads a succession of coarser or fine strata, and at each inundation entombs a new portion of the floating forest. Once more the earth is quiet, the sea becomes tranquil, holds only fine sediment and buoyant parts of vegetation. These at last precipitate themselves together by a slow subsidence and form a uniform deposit, exhibiting few traces of any horizontal currents. Thus we account for the constant reproduction of the peculiar soil of the coal seams and for the preservation of the *Stigmariæ*. Thus too we have the necessary substratum of another coal marsh. The marine savannahs again become clothed with their matting of vegetation, and are fringed on the side towards the land with vast forests of arborescent ferns and other trees; and in this way we have all the essential conditions that constituted this wonderful cycle in the statical and dynamical processes belonging to each seam of coal and the beds enclosing it. Should this theory be correct, then in opening a coal field we unclasp a whole volume of hydrographic charts, displaying, for a long succession of epochs, the ever-changing relations of the land and waters.

Whatever may be the fate of these theories, we have results in the position, form, arrangement, and even in the

F

subsequent disarrangement, of our coal fields, which are
to us of the utmost practical importance.

The natural disposition of coal in detached portions
(which frequently assume a basin-shape, and are hence
called coal-basins, and which very generally have a
tendency to an elliptical configuration), is not simply a
phenomenon of geology, but it also bears upon national
considerations, and enables us to arrive at some knowledge
of the amount of our coal. It is remarkable that this
natural disposition is that which renders the fuel most
accessible and most easily mined. Were the coal situated
at its normal geological depth, that is, supposing the
strata to be all horizontal, and undisturbed or upheaved,
it would be far below human reach, and at least several
miles under the earth's surface. Were it deposited continuously in one even superficial layer, it would have
been too readily and therefore too quickly mined, and all
the superior qualities would be wrought out, and only the
inferior left. Were it all in one uniform bed, similar
consequences would follow; but, as it now lies, it is broken
up by geological disturbances into separate portions,
each defined and limited in area,—each sufficiently accessible to bring it within man's reach and labour,—each
manageable by mechanical arrangements, and each
capable of gradual excavation, without being subject to
sudden exhaustion. Selfish plundering is partly prevented
by natural barriers; and we are warned against reckless
waste by the comparative thinness of coal seams, as well
as by the ever-augmenting difficulty of working them at
increased depths. By the separation of seams one from
another, and by varied intervals of waste sandstones and
shales, such a measured rate of mining is necessitated as
precludes us from entirely robbing posterity of the most
valuable mineral fuel, while the fuel itself is preserved
from those extended fractures and crumblings and falls,

which would certainly be the consequence of largely mining the best bituminous coal were it aggregated into one vast mass. In fact, by an evident exercise of forethought and benevolence, in the Great Author of all our blessings, our invaluable fuel has been stored up for us in deposits the most compendious, the most accessible, yet the least exhaustible, and has been locally distributed into the most convenient situations. Our 'coal fields are so many Bituminous Banks, in which there is abundance for an adequate currency; but against any sudden run upon them, nature has interposed numerous checks; while reserves of the precious fuel are always locked up in the bank cellars under the invincible protection of ponderous stone beds. It is a striking fact that, in this nineteenth century, after so long an inhabitation of the earth by man, if we take the quantities in the broad view of the whole known coal fields, so little coal has been excavated, and that there remains an abundance for a very remote posterity, even though our own best coal fields may be then worked out. This will be made evident when we come to speak of quantities and of particular coal fields.

In our own kingdom we are wonderfully favoured by the number and local distribution of our coal fields. Furthest north we see the considerable deposits of Scotland extending from the coast of Fife to the valley of the Clyde. In England, north of the Trent, we have the coal fields of Northumberland and Durham, with Cumberland and those of Yorkshire, Nottinghamshire, and Derbyshire. After these comes the large field of Lancashire, or, as it is sometimes named, the Manchester Coal field. Looking to the central districts, we see the coals of North and South Staffordshire and of Leicestershire. In the north-west we have the field of North Wales; in the more central west the deposits of the Plain of Shrewsbury, Coalbrook Dale, and the Clee

Hills ; and in the south-west we find the great coal field of South Wales, and the minor ones of the Forest of Dean, of Somersetshire, and of Gloucestershire.   By the inspection of a good geological map, we see how advantageously for commerce these several coal stores are distributed ; and they have exercised a greater influence upon the locality of men's residences than might at first be supposed.   What has made, for example, Newcastle-on-Tyne, Leeds, Manchester, Sheffield, Birmingham and Glasgow what they are, but contiguous coal fields? What has decided the locality of our vast factories and iron works, but coal fields ?   What has, therefore, determined the courses of our principal lines of railway ? primarily our great coal fields.   What has doomed some formerly populous and otherwise convenient and venerable cities to languor or decay ?   What has retarded the increase of Salisbury, Winchester, and Canterbury—all cathedral cities or towns, and all otherwise favourably situated : what but the absence and distance of coal fields ?   What, lastly, is to determine the redistribution of political power in our representative system, according to the scheme of Mr. Bright ?   Evidently the existence of coal, which attracts populations, concentrates industry, and must at length draw to the coal-bearing sites a numerical majority of the nation.

Take a geological map of a new and thinly populated country, and if it be marked by coal fields, the locality of future cities can be safely predicted from our own experience.   Men and manufactories *will* follow coal. The two former are the moveables, the latter is the fixed attraction.

Of our own coal fields, not all have been carefully examined and surveyed, yet all are known generally, and all are more or less wrought ; so that we are acquainted with their extent, larger features, and most important contents. When we are so far informed we may, by accumulating a mass of detailed observations, approach to a numerical computation of the amount of coals we possess. A survey of every coal field by some one intimately acquainted with its seams and their character, with their dislocations and denudations, and with the various geological phenomena of the whole, yet remains to be made; and the accuracy of the surveyor's knowledge of all these particulars of course determines the value of the survey. This proceeds upon a computation of the number of seams (the superficial denudation of any of them being allowed for), and their extent. The result is commonly expressed in acres with relation to particular districts, and square miles with relation to whole countries. Professor Rogers gives the total quantity of Great Britain as 5,400 square miles of coal; which nearly coincides with the latest estimates made in this country.

In these estimates nothing further is attempted than a conjecture of *coal areas*; and it is an additional step to endeavour to present an estimate of *solid contents*. This can only be done by averaging the very unequal thicknesses of all the known coal seams. Now, the total average of the thicknesses of our coal seams may be taken at from thirty-five to forty feet. This total is in most fields divisible into twenty or more beds, alternating, as we have already explained, with numerous beds of sandstone, grit, and shale. Adding together the solid contents of the coal seams of the British Islands, and assuming the average thickness of the coal to be thirty-five feet, we obtain the vast amount of 190,000,000,000 tons of coal. Respecting all such estimates it must be remarked that, however carefully calculated, they do not represent the amounts of coal actually available to us, but only those which are supposed to be deposited in the coal fields.

Amongst our own coal basins none can compare in importance and notoriety with the great northern one, shipping its coals by the three rivers—the Tyne, the Wear, and the Tees; or, as it is commonly termed, the Newcastle coal field. It has long been famous as the source of the best household fuel for our own country, and also for no small part of Europe. We shall, therefore, select it as the best example for detailed consideration, omitting any lengthened notice of other British deposits, except the Preston and Welsh coal fields. It is bounded on the north by the river Coquet, and extends southward nearly as far as Hartlepool, on the river Tees, that is a distance of about 48 miles. Its extreme breadth is about 24 miles, and its area has been stated as 800 square miles; but more careful measurements would leave us to adopt the calculation which affords little more than an area of 700 square miles. The lowest working seams are at Monkwearmouth, near Sunderland, where the celebrated shaft of Pemberton's pit has been sunk 1,710 feet before the Hutton seam of coal was reached; a depth which is feebly conceived by a mere inspection of figures, but which can be strikingly represented by saying that this shaft exceeds in depth the Monument of London piled eight times upon itself, and nearly equals St. Paul's piled upon itself five times! This for a long period was regarded as the deepest perpendicular shaft in the world, but some foreign mining shafts now dispute with it the palm of depth, and a shaft near Ashton-under-Lyne approaches to it. As a mining enterprise, however, and a mining triumph it has not its superior in any locality. Undertaken by a few private individuals, in ignorance of the depth of the coal, they might well have been daunted by its apparently hopeless nature, and by the fearful expenses incurred.

Ill prophecies were not wanting,—blacker than the coal they could not reach—and many professional opinions were against them. Hopeful, fearless, and unusually enterprising, they persevered; at a cost, it is said, of little less than 80,000*l.* Success crowned their efforts, and that valuable coal, the Hutton seam, was won.

A very important question has from time to time been discussed with particular reference to this great coal field, viz., is the present rate of excavation such as to lead to any well-founded apprehension of a scarcity of coal within a period not very remote? A reply to this question can only be afforded when we have something like a true knowledge of the real areas, number, and solid contents of the several coal seams, and also of the present produce of the existing mines. It is in the consideration of this enquiry, so interesting to all, that even the minutest details of the mining of our best coal becomes valuable. Former estimates have been found baseless for lack of accurate elements of computation. Thus, in 1830, Dr. Buckland limited the future supply, at the then existing rate of waste of small coal and consumption, to a period of 400 years. He was nearly right upon wrong data, for he erroneously thought that there was no coal beneath the limestone, a once prevalent opinion. Another estimate, founded upon the assumption of an area of 837 square miles of coal in this district, and an already excavated portion of 150 square miles, protracts our lease to 1,700 years to come. Other calculations give us a future of 1184 years, or, as a minimum, 1000 years.[1] We

---

[1] The prophets of our prospective impoverishment in coal perhaps deserve distinct mention. Their prophecies are curious, if not correct. In 1801, Mr. Bailey predicted the supply of the Durham and Northumberland fields would only last 200 years; in 1792, Dr. M'Nab named 375 years; in 1830, Dr. Buckland granted us 400 years; in 1830, Professor Thomson extended our good fortune to 1000 years; and in 1830, Mr. Hugh Taylor exceeded all previous calculators by granting 1727 years.

believe, however, that not more than one or two enquirers
have thoroughly investigated this matter in connection
with our present knowledge of details. We now learn
from the observations of local colliery managers that there
are about 57 different seams of coal in the Great Northern
or Newcastle field. These vary in thickness from 1 inch
to 5 feet 5 inches and 6 feet, and they form an aggregate
of about 76 feet of coal. Some, however, of these seams
being so thin would not be remunerative to work, at least
at present. Experience proves that, to defray the cost of
mining, and to have a fair return for risk and interest of
capital, no seam can be worked, at any considerable
depth, which is less than two feet in thickness. Three
estimates have been made of the saleable coal acreage of
this field; and they are as follow :—

Estimate of Mr. Hugh Taylor .   .   . 535,680 acres.
  „   Mr. R. C. Taylor .   .   . 499,200   „
  „   Mr. T. Y. Hall   .   .   . 471,680   „[1]

We have only to ascertain the annual extraction from
this field in order to conjecture its duration to us. Now
in 1854 the annual produce of coal from Northumberland
and Durham was 16,221,001 tons. Abate this to an

---

[1] In another shape, the quantities contained in the ten principal seams
of this coal field have been computed by Mr. Greenwell separately, and he
finds the result to be as follows, in Newcastle chaldrons (53 cwt.):

	Chaldrons.
Gross chaldrons in the whole	2,180,551,561
Deduct quantity supposed already extracted .	303,702,805
Quantity now remaining in the field . .	1,876,848,756
Deduct loss by underground waste, ⅙ . .	312,808,126
Quantity remaining to be raised below surface .	1,564,040,630
Deduct loss by present mode of screening, say one fifth	312,808,126
Total merchantable round coals . . .	1,251,232,504

annual average of 10,000.000 tons (it will, in all probability, be half as much more), or 3,773,585 Newcastle chaldrons of round coal; and then we can easily reckon that the reduced amount will be exhausted in 331 years. Should the demand and supply increase, as they have done of late years, we may affirm, ín round numbers, that three centuries will see this great coal field exhausted or hopelessly impoverished.   To confirm the authority of the figures, it is remarkable that a mining engineer (Mr. Greenwall) arrived, in 1846, at the same result as Mr. Hall has arrived at more recently, namely, 331 years. Both these gentlemen have assured us that after trying various methods of computation, they have come to very nearly the same final figures. We have only given the results, but having looked over the details of each seam, we are disposed fully to rely upon their statements.

Some of our other principal coal fields might be the subject of similar prophecies.   For instance, the immense consumption of coal in the iron furnaces and foundries of Staffordshire will probably lead to an exhaustion of that coal field even before Northumberland and Durham; for its area is scarcely more than one half of the area of the Northern coal field.   It has, indeed, one very thick seam of coal of from thirty to forty feet; but this will not alone counterbalance the difference.   Wherever coal fields are situated amongst numerous iron works and manufactories, as well as large populations, there is a continual and increasing demand upon the produce of the mines; and thus even Yorkshire, Lancashire, and Derbyshire are more than living up to their income of coals.   The quantity for supply being fixed, and the quantity demanded being continually on the increase, the actual period of exhaustion is not difficult to predict, though it may be unwelcome to anticipate.

We shall, however, have resources at that period which

will prevent a bituminous bankruptcy; and these will be
found in one or more coal fields not at present so largely
worked as those previously named : particularly the great
coal field of South Wales will afford an abundant supply
for many years to come. We prefer to speak in detail of
this field when we come to treat of steam coal, of which it
is the great repository, not only for ourselves but also for
foreign markets.

In all departments of activity and ingenuity the advance
has been commensurate with the demand and supply of
coal. Indeed, we doubt, whether any equal space in the
world, not occupied as building ground for towns and
cities, has witnessed such a rapid development of human
labour and resources. Few sights of a commercial kind
are more impressive than those which may be every day
and night witnessed in these districts, where coal-waggons
are careering in successive trains over far-stretching railways, and hurrying down to rivers and ocean, until they
are unloaded, and their contents shipped by gigantic
machinery. Steam engines are unceasingly at work
drawing coals and pumping out water. Thousands of
men are underneath our feet cutting down the coal by
severe and peculiar labour. Thousands are around us
receiving loads and despatching them by railways, and
screening the coals by dashing them upon huge screens
standing in long rows, whence fly up black clouds of
impalpable coal dust, filling ears and eyes and throats
with microscopic specimens of coal.

Go where you will, there is a network of small railways
leading from pit to pit in hopeless intricacy, but all having
a common terminus on the river's bank, or the ocean's
shore. Go where you will, tall chimneys rise up before
you; and here and there a low line of black sheds,
flanked by chimneys of aspiring altitude, indicates that
you are arriving at a colliery. . As you draw nearer, men

and boys of the blackest hue pass you and peer at you
with enquiring glances. Now trains of coal waggons rush
by more frequently, noises of the most discordant character increase, and you know that you are at the pit's
mouth when you behold two gigantic wooden arms slant-
ing upwards, upon which are mounted the pulleys and
wheels that carry the huge flat wire ropes of the shaft.
For a moment the wheels do not revolve—no load is
ascending or descending—but the next minute they turn
rapidly, and up comes the load of coals, or human beings,
to the surface.   Perhaps the most impressive sight is a
large colliery fully engaged at night work, with burning
crates of coal suspended all around; and after this a view,
from some neighbouring eminence, of all the far-flaming
waste coal heaps, burning up the accumulation of waste
and small coal not worth carriage, ever added to the ever-
consuming mound, until the whole district appears like the
active crater of some enormous volcano.

It is difficult to form anything approaching to a correct
estimate of the produce, the destination, and the consumption of coal; but if 16,000,000 tons be the present annual
yield from the Northern coal field, we are assured that the
total annual produce of our 3,000 British coal mines is
(as the maximum) no less than 68,000,000 *of tons!*—a
quantity more than double of what had been conjectured,
but now confirmed by careful researches and unquestion-
able authority.   It is very difficult to convey an adequate
conception of this vast produce ; but if, as a collier has
calculated, these sixty-eight millions of tons were excavated from a pit-gallery six feet high and twelve feet
wide, such a gallery must be 5128 miles and 1090 yards
in length.   Or if, instead of this tunnel of more than
five thousand miles, we prefer the conception of a solid
globe, then the diameter of a globe containing this annual
produce must be 1549·9 feet.   Should a pyramidal form

be chosen, then this quantity would constitute a pyramid, the square base of which would extend forty acres, and the height of which would be 3,356·914 feet.[1]

It has been doubted whether it be possible to form any adequate conjecture of the extent and amount of coal in the principal countries of the world.[2] But although geo-

[1] The annexed table has been prepared to show, in a compendious form, the aggregate distribution of coal over the United Kingdom, in one recent year, 1854. In the year 1858 the total produce was rather less, as previously stated.

Counties and Districts	Annual Yield	Quantities Shipped	Used for Iron Working	Used for Sundries
	Tons	Tons	Tons	Tons
1. Durham and North- umberland . .	16,221,001	8,688,551	4,300,000	3,232,450
2. Cumberland, York- shire, Derbyshire; in all seven coun- ties . . .	15,811,670	719,913	3,862,780	11,228,977
3. Eight counties .	16,189,366	587,000	3,734,693	11,867,673
4. North and South Wales, Scotland, and Ireland .	17,239,750	8,824,047	10,650,000	2,765,703
Totals . .	65,461,787	18,819,511	22,547,373	29,094,803

From these totals it will be seen how large a proportion of our annual yield of coal (more than one third of the whole) is absorbed by iron works. This, again, is chiefly appropriated to a limited area, where iron works are situated. Thus we find that in 1854 Staffordshire and Worcestershire yielded 7,500,000 tons of coal, out of which they consumed for their manufacture of pig iron alone, 3,415,200 tons. So also, six Welsh counties unitedly produced nearly 10,000 tons of coal, and consumed more than 5000 tons in their iron works. In Scotland nearly five sevenths of the annual coal produce was applied in the same way.

[2] It has been found impossible to arrive at true statistics of coal in almost any country, nor was our own an exception until very recently. Mr. Richard C. Taylor was the first who made the attempt, and he published in the United States, a laboriously compiled volume entitled 'The Statistics of Coal,' which we reviewed in the XCth volume of this journal, page 525, and we refer our readers to that article for an abridgement of Mr. Taylor's interesting calculations and diagrams. The first edition appeared in 1845, and a second edition, somewhat enlarged by others, appeared in 1855.

logical surveys of many foreign coal-fields remain to be made, and authoritative statistics are very imperfect, yet enough has been done by geologists, and by public and personal research, to enable us to arrive, at least, at an approximative estimate of this nature. As Professor Rogers has not hesitated to put forth an estimate of American and European coal fields, we will adopt it, and arrange it for our present purpose in à note. Doubtless advancing knowledge will lead to an amendment of some figures, nor are we confident that some of them should not be reconsidered; but admitting such possible defects, the statement is highly interesting and suggestive.[1]

[1] *Summary view of American and European Coal Fields.*

The aggregate space underlaid by the vast coal fields of North America amounts to nearly 200,000 square miles or to more than 20 times the area, including all the known coal deposits of Europe, or indeed of the whole Eastern continent.

Comparing the assumed areas and solid contents of the coal fields of other countries with those of North America, we have the following results:—

I. *Estimated Areas of Coal in Principal Countries.*

			Total Square Miles
United States . .	196,650 square miles of coal area		200,000
British Provinces of N. America . .	7,530	ditto	
Great Britain . .	5,400	,,	
France . . .	984	,,	
Belgium . . .	510	,,	
Rhenish Prussia, Saarbrücker coal field .	960	,,	
Westphalia . .	380	,,	8,964
Bohemia . . .	400	,,	
Saxony . . .	30	,,	
Spain . . .	200 (?)	,,	
Russia . . .	100	,,	

II. *Estimated Quantities of Coal in Principal Countries.*

	Tons
Belgium (average thickness, 60 feet of coal) .	36,000,000,000
France (about the same thickness) . . .	59,000,000,000

It will be perceived at once that no characteristic of the northern continent of America is more remarkable than the unbounded fields of coal which it possesses. Nearly 200,000 square miles of coal fields can scarcely be grasped at first thought; and if we should go back to the growth and accumulation of vegetable matter necessary to their formation, can anything that we now behold on the surface of the globe afford us a parallel in an equal space? Our own coal fields, in the aggregate, would form but a black speck beside them, upon any map. The possession of such an amazing deposit leads us to forecast a future of almost boundless enterprise and production for that wonderful country. We must however wait, before we prophesy, to learn more of the character of the North American coal. At present it has been but little worked compared with its extent, and what has

	Tons
British Islands (average thickness, 35 feet) . .	190,000,000,000
Pennsylvania (average thickness, 25 feet) . .	316,400,000,000
Great Appalachian coal field (same thickness) .	1,387,500,000,000
Indiana, Illinois, Western Kentucky (25 feet) .	1,277,500,000,000
Missouri and Arkansas Basin (10 feet in thickness)	739,000,000,000
All the productive coal fields of North America (assuming thickness of 20 feet of coal over 200,000 square miles)	4,000,000,000,000

III. The *Ratio of the estimated quantities of coal* in the more important of these several coal countries is shown approximatively in the following series of numbers, making the coal of Belgium, or 36,000,000,000 tons, our unit of measure :—

Amount of coal in Belgium	. . . . . .	1
,, France, less than	. . . .	2
,, British Islands, rather more than	. .	5
,, Pennsylvania, a little less than .	. .	9
,, Appalachian coal field, about	. .	38½
,, Illinois, Indiana, Western Kentucky Basin		35½
,, Missouri and Arkansas Basin	. .	20½
,, Entire coal fields of North America .	.	111
,, ,, of all Europe .	. .	8¾

been brought to market cannot be regarded as equal to our own best household coal. The total produce of America, for the whole thirty-five years from 1820 to 1855, did not exceed the produce of Northumberland and Durham for four years, from 1851 to 1855, while it was less than the total annual yield of the United Kingdom by 7,000,000 tons. The total United States' produce, in 1855, was 7,600,000 tons; and the same amount was, in the same year, yielded by Scotland alone.[1]

That the Americans have not yet derived full benefit from their extraordinary coal deposits, may be supposed; but it is not commonly known that, owing to the distance

---

[1] The great coal deposit of Pennsylvania (though itself comparatively small) is anthracite, and the immense coal fields in the valley of the Mississippi are composed of slatecoal, which is of a similar character.

It has been observed by Professor Rogers (who personally pointed this out to us, upon an unpublished geological map of America,) that there is a geological feature of high interest, connected with the position of the comparatively small anthracitic basins of America. They lie bordering upon a long ridge of contorted strata, highly metamorphosed, and the metamorphic action seems to have so affected the adjacent coal, as to convert it into anthracite, by burning out the bituminous constituents, and so far coking the coal, which becomes less and less coke-like or anthracitic in proportion as it recedes from the great ridge of contorted rocks, and finally, far away, is found pure bituminous coal, in great natural basins. The great hilly ridge has acted like a lengthened range of fire, partly coking the coal nearest to it, and exciting less and less influence in the ratio of distance. Nor is there any igneous rock near to explain this action upon the acknowledged principle elsewhere observed; and the whole effect is attributed by the professor to volcanic heat, steaming up from old volcanoes, and causing the observed phenomena. From a similar cause, in comparatively recent earthquakes, near the Mississippi, it was found that a long line of the snow, then lying upon the ground, had been melted, while the mass of snow was unaffected.

We believe that we can find an analogous phenomenon (if not due to the same cause) in our own South Welsh Coal Field, where the change from bituminous coal to anthracite seems to have been produced by the intense lateral pressures to which it was subjected when the long ridges dipping in opposite directions, like the roof of a house, and geologically termed *anticlinals*, which bound and intersect it, were originally formed.

of their great coal fields from the centres of residence and commerce, and the absence of railroads and canals, and all but the most costly means of carriage, together with the situations of some of the coal basins behind mountains, the country has but little beneficial advantage at present in its rich natural possession. Last year the comparatively limited coal fields of Great Britain exported 363,628 tons of coal to North America, of which the United States received at ports of the Atlantic, 284,869 tons. These issued chiefly from our northern ports and from Liverpool; the former shipping 115,147 tons. These figures prepare us for the fact that a considerable proportion of the coal consumed in the houses of New York, Boston, and other chief cities of the United States, is British, and that at a cost to the consumer very consolatory to the exporters. We are informed that the English coal going from Liverpool to American factories costs about 2l. per ton there. The most luxurious classes demand our Cannel coals (a Lancashire product), which burns with a fine clear flame, whence its name of candle or cannel coal, and this costs, in New York and Boston, nearly 4l. per ton. An American invalid assured us, that having to burn a certain proportion of cannel coal, his annual charges for coal, in a moderate-sized house at Boston, amounted to 50l.; upon removing to reside in a British coal country, he found that he saved nearly 40l. a year in coals. The general custom in American cities is to burn anthracite, which can be delivered at New York at prices varying from 16s. per ton and upwards, in large stoves placed in the cellar, from which regulating pipes convey the heat to different parts of the house. Those who cannot bear this dry, unwholesome heating, burn, in addition, good bituminous coal, probably English, in the open grate, and thus the heat radiating from the open grate, in some degree counteracts the dryness of the stove warmth.

The reports of the results of experiments carried on for Government in relation to supplying coal for the steam navy, greatly favours the South Welsh steam coals. The consequence has been a rapid development of the produce of this coal field, and such an increased interest in its extent and products as will justify a brief notice of both.

The area of the South Wales coal field has been determined, from computations upon the ordnance and other geological maps, to be about 1055 square miles. It commences on the east in the county of Monmouth, and extends westward through Glamorganshire, Brecknockshire, and Carmarthenshire, to the western boundary of the county of Pembroke. It lies parallel with the Bristol Channel, the numerous harbours of which afford great facilities for exportation, and the railways now constructed in the principality will have the effect of diminishing the expense of carrying the coal from the collieries to the shipping ports and markets for home consumption. The coal is distributed over the included counties, thus:—In Monmouth it is found in the western part of the county, abutting on Glamorganshire in the north-west, and on Brecknockshire in the north. Here its length is about seven miles, and its breadth fifteen and a half miles; the whole area being 110 square miles, and the coal itself bituminous and free burning. It is extensively employed in the smelting and manufacture of iron, at the Tredegar and other large works on the northern boundary, while considerable quantities are shipped at Newport. Four-fifths of the entire county of Glamorgan are covered by coal. This area contains every kind and quality of coal found in South Wales, and embraces an extent of 546 square miles; of which 462 square miles are bituminous and free coal, and eighty-four miles are anthracitous coal. Out of the above-named

G

amount twenty-six square miles lie under the sea in
Swansea Bay. Large quantities of the bituminous coal
are used in copper smelting, and in the manufacture of
tin plates. In Brecknockshire there are only seventy-
eight square miles of coal, which are principally situated
on the south-west corner, adjoining the counties of Glamorgan and Carmarthen. The whole is called anthracitous. Carmarthenshire possesses 241 square miles
of coal, 105 of which run under the sea in Carmarthen
Bay. Of this entire district, 117 square miles are styled
bituminous, and 124 square miles anthracitous. Both in
this and the previous county much of the anthracitous
coal is employed in the smelting and manufacture of iron,
and a considerable amount is shipped at Llanelly. Passing on, the coal continues westward through the whole
length of Pembrokeshire, in a strip which narrows as it
proceeds, to St. Bride's Bay, at the extremity of the
county; and there it holds on its course along the bay,
at right angles to its former direction. Its whole area in
this county is but eighty square miles, and the whole is
anthracite, of which some is of superior quality. It is
preferred to bituminous fuel in the agricultural counties,
to which it is exported, and where it is used in stoves, for
drying hops and malt, and burning lime. Thus, out of
this whole field of 1055 square miles, we have 689 square
miles of bituminous, and 366 square miles of anthracitous
coal, as respects coal superficies or area. When we
spoke, some pages back, of prospective impoverishment
in the North of England, we alluded to this great store,
and here we have the counterbalancing prospective abundance—in quantity—though not in equal quality. The
present returns from all the collieries of South Wales
under government inspection, give, for 1858, the amount
of 7,495,289 tons, not quite half the produce of Durham
and Northumberland for the same year.

But our chief interest, at present, is in the steam coal of South Wales. Highly bituminous coal not only produces a small percentage of coke, on account of the large proportion of volatile matter which it contains, but its quality is much inferior to that produced from kinds less bituminous. Being, also, specifically light and spongy, and much honeycombed, it is soft, easily crumbles, and therefore greatly wastes. The coal named semi-bituminous, however, is very valuable, and is suitable for almost every purpose except making gas. Sufficiently bituminous to be easily kindled, it makes a bright cheering fire, and gives out great heat with very little smoke. It is also important to notice that its smoke, instead of being black and dense in large volumes, as in the highly bituminous coal, is inconsiderable in quantity, brown in colour, and productive of little soot. The anthracite of South Wales burns without emitting flame or smoke, does not ' bind ' or cake, or soil when handled, and has a general metallic lustre. Its ash is of a light pink and sometimes of a dark grey colour. Properly speaking, it does not form a coke in the usual acceptation of that term ; for the water and hydrogen are expelled in small quantities during the distillation, and a slight diminution of bulk takes place, yet no new arrangement, as is usual in the transition of coal into coke, is formed ; the fracture remains the same, and there is not that cellular structure which every one may observe in common coke.

The employment of anthracite in our steam-ships is a subject of great importance ; and if it should be found that it can be successfully adopted, part of the space now occupied by the stowage of coal might be saved ; and appropriated to an increased cargo, as anthracite, bulk for bulk, is of the greatest density of any coal. Taking into account the greater specific gravity of anthracite, a saving of one third of the space now occupied by coal might be saved; and it has been proved that 100 tons of anthracite will do the duty of 144 tons of bituminous coal. It is probable that by the use of the best and best-picked anthracite, the necessity for calling during long voyages at coaling stations might be in a great measure avoided and in the event of a naval war, such provision for fuel might be made as to give a decided superiority to the steam-ships so furnished. At present, during the year 1858, the anthracite district of South Wales yielded 737,590 tons. Anthracite might, we believe, be rendered serviceable for locomotive engines by admixture with other coals, but its decrepitation, or flying to pieces when heated, will, we fear, prevent its exclusive adoption in this way. It was, indeed, tried in the engines on the Liverpool and Manchester Railway about 1839, but the draught up the chimney was so strong that the coal was projected into the air in fine powder, and the carriages were covered with it.

The American Government have been more alive to the importance of possessing proper surveys of their anthracitic deposits than our own. Professor Rogers was particularly instructed to accomplish this task with especial relation to commerce, and one of the tables in his book shows the development of the Pennsylvanian anthracite mines from the commencement, in 1820, through all the stages of growth. In 38 years the trade advanced at the rate of 184,000 tons per annum, and from 1839 to 1849, the produce doubled itself in each five years, while it has again doubled itself in eight years, so as to attain in 1857 the aggregate of 6,431,378 tons of hard anthracite. Vast as our British coal trade is, it has only doubled itself in about twenty-four years. Anthracite is extensively used in the States as a manufacturing coal. For iron-smelting and iron-melting in the foundry, it has been employed during the last dozen years, and the iron works have been carefully adapted to its use. We learn, also, that it is in extensive employment as a steam coal in the steam-boats of the American rivers, and in the American lines of Atlantic steamers, as well as in the steam navy of the States. In fact, this kind of coal is now generally employed in the United States for most purposes for which a mineral fuel is required.

If we examine the little that has been reported respecting the coal fields of France, Belgium, Germany, and Russia, we do not learn that as yet any considerable amount of good anthracite or steam coal has been discovered in those great countries. The mean annual produce of coal in France, deduced from returns for five years, is only 5,490,702 English tons (of 2,240 lbs.); that is, about the thirteenth part of British produce at present. In five years France has not extracted the half of our last year's amount, and in five years she has produced of anthracite only 3,597,220 tons. It is doubtful whether her anthracite forms a good steam coal, while the experiments on fuel which have been made in the French Navy have demonstrated the superiority of British coal over the produce of the coal fields of France. An attempt has obviously been made to provide against the evil day by accumulating large stores of British coal in the French ports. Last year we exported to France 1,344,342 tons of coal, of which 354,364 tons issued from ports on the Severn, and the remainder from Northern ports. But the effect of a prohibition of the export of coal during hostilities would be speedily to exhaust the stores of this essential combustible, and not only to embarrass the operations of foreign navies, but to interrupt to a very great extent the manufacturing power and the supply of coal gas on the continent.

In conclusion we must remark that in the official and

concerted survey of their mineral fuels the Americans have surpassed us. When we look at such a work as this of Professor Rogers,—laborious in statistics, accurate in surveys, and magnificent in form and embellishments; when we bear in mind that this is merely the survey of one State and mainly the results of one man's personal labours and studies; we may well turn and enquire of our own scientific authorities what we possess of a similar character? The answer must be humiliating: we have simply nothing worth a moment's comparison. We are the first coal-mining and coal-producing country in the world. If we assume the entire annual coal produce of the chief coal fields in the world to be 100,000,000 tons, we ourselves contribute more than three-fifths of that quantity, and the estimated money value of our annual coal produce amounts to the amazing sum of *sixteen millions and a quarter*. We have deposits of the most varied character and the most valuable qualities; we have a very far larger amount of capital invested in coal-mining than any other nation; we employ above 200,000 persons of all ages in this kind of labour; we have some of the deepest and largest mines, and the most stupendous accumulations of steam power for pumping out water and drawing up coals; we have the most expeditious and ingenious methods of shipping the produce; we have at this hour in our coal-mining districts, scenes of activity above ground, and galleries of mining industry under ground, which astonish all foreigners who care to glance at the one and dare to descend to the other; we have mining engineers of large experience, and even wealth and social position; and we have a national stake in the whole of at least as great importance as we hold in any department of British industry; but we have no adequate publication on the subject—we have as yet no complete surveys of our coal fields—no uniform and official maps of the whole—no compact and continuous account of their mineral character and contents. Here are Professor Rogers' three beautifully illustrated quarto volumes on one American State, and we have not three illustrated quartos on the whole of our British coal fields. It is only a few years ago that we learned what our annual produce of coal really was, and it was then found to be so much in excess of what had been previously conjectured as to appear incredible. At this very time, with the exception of mere statistics officially published, we have no means of tracing some of the most interesting and important circumstances connected with supply and demand. That these are facts, no one can deny; that our ignorance is indefensible, every impartial enquirer will acknowledge.[1]

[1] Since the above lines were written, the Geological Survey has industriously progressed. At this date (May 12, 1873) I ascertain from official sources that the following is the condition of the survey in relation to our coal fields :—

Geological Survey of English Coal Fields.

Northumberland } Almost surveyed;
Durham   . } some maps published.
Yorkshire   . } Completely surveyed,
Lancashire . } but not all published.

The survey of these coal fields is completed, but the whole of the maps are not yet published.

The Whitehaven coal field is being surveyed.

All the other important coal fields are published on a one-inch scale, and will eventually be published on a scale of six inches to the mile.

III.

FATAL ACCIDENTS IN COAL MINES.[1]

THE TITLE of the Parliamentary volume which heads
this article is in part a misnomer, for it contains no
report, but merely a large amount of desultory and miscellaneous evidence prefaced by an intimation that the
committee propose to resume their labours in the present
session. The primary purpose of this committee was to
enquire into the operation of the Acts for the Regulation
and Inspection of Mines, and into the complaints contained in petitions from some 14,000 coal-miners of Great
Britain, which had been presented to the House of Commons. The desires of the petitioners, as expressed in the
terms of a ' Petition of the Under-Miners of Northumberland and Durham' given in an Appendix, appear to be
reasonable, but the justice of their complaints and the truth
of their statements must, of course, be tested by evidence.

These grievances relate to several subjects connected
with coal-mining—such as the modes of estimating work
done and paying wages, but the most momentous complaints are those relating to ' the fearful sacrifice of life in
mines and collieries,' which, say the miners, ' affords
abundant proof that the legislative measures hitherto
passed have proved to be totally inadequate for securing

[1] I. *Report from the Select Committee on Mines; together with the Pro-
ceedings of the Committee, Minutes of Evidence, and Appendix.* July 1866.

II. *Coal Mines (Accidents and Explosions).* Return of a Copy of a Circular
Letter from the Home Office to, and Reports from, the Inspectors of
Mines to the Secretary of State for the Home Department on the recent
Accidents and Explosions in Coal Mines, &c. February, 1867.

the personal safety of the miners of this country.' To
this matter alone shall we direct the attention of our
readers, with a view of putting them in possession of
such information as may qualify them to form an opinion
upon a subject which has recently awakened universal
public interest, in consequence of the late terrible explosions in Barnsley and in Staffordshire.[1]

The public are by no means aware of the actual loss of
life occasioned by coal-extraction in this country. By
searching into various local publications in the North of
England, and by a fair estimate of probabilities arising
from what has been discovered, we are quite warranted
in assuming the total number of lives sacrificed in our
coal-mining, from the earliest notices to the year 1850, to
be not less than 10,000. This is certainly not too high
an estimate, and probably a very low one. In November
of the year 1850 the first Act for the Inspection of Coal
Mines came into operation, and henceforth we have some
authentic data for accidents. During the ten years from
1850 to 1860, the deaths in or at all the British coal
mines amounted to 9,090. In the ensuing five years,
ending 1865, the deaths were altogether 4,827. Thus
then adding to the 10,000 deaths up to 1850, 10,000
more (in round numbers) up to 1860, and nearly 5,000

[1] In the present article we are dealing exclusively with accidents termi-
nating fatally. The impression which these will make might be very much
deepened by an estimate of the additional number of serious but not fatal
accidents. These are very numerous, but there is no return of them. Mr.
Dickinson, inspector for the Manchester district, observes, 'There are about
sixty fatal accidents every year in my district, and besides these fatal
accidents a large number of non-fatal accidents that require investigation.'
Of course these are quite unknown to the public. During a visit of some
months paid to one of our principal coal-fields, we had daily opportunities
of seeing and conversing with the victims of such accidents. We conversed
with many pit-lads who had been 'lamed' (or injured) several times in a
few years, and who reckoned events by the chronology of their various
'lamings.'

more up to the close of 1865, we have in all an estimate
of nearly 25,000 *deaths from coal-mine accidents,* from the
commencement of any account of them to within little
more than a year of the present date.

We apprehend that the melancholy mortality which we
have just estimated has never before been brought, as a
whole, before the country; and certainly it has never
been sufficiently considered in its full and aggravated
interest.   Twenty-five thousand persons have been
snatched from our industrial population in the midst of
their occupations, and not only so much human life, but
likewise so much skilled labour has been removed from
us.   Not the infirm and useless, but the able and the
industrious, have been thus hurried away.   No other
kind of work is attended with so many and so fatal
accidents ; and if there were suitable data for the comparison, it could probably be shown that our various
wars of recent date have not deprived us of a greater
number of human beings, while all the deaths in and
around mines have taken and are taking place in periods
of profound peace and high prosperity.

A complete list of deaths from colliery accidents for
the ten years commencing January 1st, 1856, and ending
December 31st, 1865, has been compiled by one of the
present inspectors, Mr. Atkinson, of which the following
table is a summary :—

Causes of Death	Number of Deaths	Proportion Per Cent.	Amounting to
Deaths resulting from fire-damp explosions	2,019	20·36	About one-fifth
Deaths resulting from falls of roof and coals	3,953	39·87	About two-fifths
Deaths resulting from shaft accidents	1,710	17·24	Less than one-fifth
Deaths resulting from miscellaneous causes and above ground	2,234	22·53	More than one-fifth
	9,916	100·00	

We now proceed to notice the four principal causes of these mining deaths, taking the above summary as our text.

First and foremost come the Explosions of Fire-damp, which, though not so frequent in occurrence as some others, are certainly the most sweepingly destructive and fearfully interesting of all colliery calamities, and, at the same time, the most difficult to be guarded against—and those most exacting an unremitting vigilance and experimental knowledge. We must presume that our readers are acquainted with some elementary principles in connection with these catastrophes. Coal-pits are divisible into two large classes—gaseous and non-gaseous, or, as the miners say, ' fiery ' and non-fiery pits. The former might again be subdivided into fiery and very fiery pits, in accordance with the less or greater amount of gaseous emanations. Fire-damp, or light carburetted hydrogen gas, exudes from the coal containing it so soon as that coal is worked, and continues to exude at intervals and in proportion to the amount of atmospheric pressure. There seems much reason to suppose that it exists naturally in a high state of tension in the coal-beds, and that when the pressure of the atmosphere decreases more gas escapes. The common air, in fact, to a certain extent, weighs upon and imprisons it. In some very gaseous coal its escape is audible in a low hissing sound, which we have sometimes listened to in the gloomy stillness of a dangerous mine. Unhappily the best burning coal often produces the most gas, as is the case in the old and only true Wallsend pit near Newcastle. That very fiery mine is now closed, and probably closed for ever. Well is it that it should be closed, for such a seething gasometer is there covered up, that a mere slender tube let down to it and brought from it to the open air, for a long time sent forth an unfailing stream of gas which being lit would flame

night and day continuously. There is gas enough in the
broken wastes of that famous old mine to light up the
whole town of Newcastle, and probably to blow the town
into ruins if exploded underneath it.

To render this gas actually explosive,[1] it must be mixed
with at least four times its volume of atmospheric air.
When mixed with less than four times, or more than
sixteen times its volume of air, it will not explode. Shut
out the air from it altogether and it is harmless, give it a
little air and it becomes dangerous, a little more and it is
more dangerous, and with still more air it becomes highly
explosive, until a large addition of air deprives it of all
its devastating power. Hence a gaseous pit but slightly
ventilated is thereby brought into the very condition of
explosiveness, and the only effectual way of dealing with
it is to dilute the gas with an abundant influx of pure
air. Let it be mingled with common air to twenty or
thirty times its own volume, and then it may be led like
an enfeebled and fettered enemy along the mazy air-
ways of the pit, and finally be dismissed to the clouds
above.

To do this is the aim of Ventilation, a word which,
while it suggests but a simple process to the casual reader,
calls up before the view of a mining engineer a series of

---

[1] Here we use terms popularly, for 'fire-damp' is a name given to different combinations of gases, the only combustible element in which is the
carburetted hydrogen. As may be supposed, the vulgar names for several
mine-gases are not used with any distinct knowledge of their real properties
or differences. Chemically speaking, fire-damp, or marsh or mine gas, is
composed of one equivalent of carbon and two equivalents of hydrogen.
Its specific gravity is nearly half that of common air, and by its levity it
tends to rise to the upper parts of passages. On its combustion it forms
carbonic acid and water. Carbonic acid is thus generated by a mine explosion, and is so dense as to accumulate at the floors of passages. Being
poisonous it becomes fatal to human existence, and suffocates those who
inhale it. Further on we shall speak of this 'after-damp' more particularly.

extended and carefully devised arrangements which, at least in the North of England, have attained to a complex and highly artificial system. The principle of the arrangement for ventilating a coal-pit is that of two distinct shafts, a downcast and an upcast. At the base of the latter a huge fire of coals is kindled, and the effective ventilation is measurable by the difference of the weight of the two columns of air in the two shafts. The lighter the column in the upcast, or the greater the rarefaction, the better is the ventilation. Yet there is a practical limit to its amount, which has been much discussed amongst pit engineers and men of science. We cannot here explain what some men of science have called the Furnace-Paradox; but the substance of what they affirm is, that the effective power of the furnace is sooner reached than was once expected, and that beyond that limit no consumption of coal can produce increased results. Therefore, in lieu of the furnace, Mr. Gurney proposed to substitute the employment of a number of jets of steam, the emission of which at the upcast shaft would produce, as he alleged, a more powerful draught and a less irregular and uncertain current than that obtained by the furnace. This plan was fully considered at the time of its proposal, and shortly afterwards some of the eminent mining engineers of the North made experiments with steam-jets, and compared their effects with those of the furnace. Having perused the details of those experiments, we are compelled to conclude that, ineffective as the furnace is beyond a certain limit, the steam-jet cannot be recommended as an advantageous substitute for it.

The reader will at once perceive that the chief essential is a ventilating draught which will not reach its limit before it succeeds in sweeping out as with the strong wings of a steady wind the noxious emanations of the mine. In an extensive mine, where the passages are long, numerous, tortuous, and small, and where there are many human beings at work vitiating the air, it must be a very powerful impulse which will drive an adequate air-current, in good and pure condition, through several miles of subterranean excavations, and carry it over all obstacles to its final discharge into the upper air. Add to these obstacles the natural *drag* or retardation by friction, which gradually weakens the air-current, so as to make what is at first like a tight rope become at last a slack one, and the necessity for great power of impulse is still more apparent. An air-current in a large pit may exceed ten or twenty miles in its entire length, and, like a traveller bound on a benevolent mission, it must retain its energy undiminished to the last, and give out its refreshing succours at every intermediate station.

The furnace system has long prevailed in British coal mines, but it has been largely abandoned in those of Belgium ; and a Belgium engineer who lately visited us declared that we were ten years behind his country in coal-mining engineering, particularly instancing our retention of the furnace. Several minor evils attend the use of it, one of which is that it speedily injures the mining gear in the shaft which it heats. In some instances the men descend and ascend by the heated shaft, and it may readily be conceived how injurious it is to go down and come up a long subterranean chimney, which a furnace shaft really is, with a monster fire burning at its base. Some explosions have been attributed to the furnace, at which the ' return air,' loaded with gas, has been said to have ignited. The Belgians use and advocate mechanical fans, which act somewhat in the manner of huge paddle-wheels in steam-ships, and by rapid rotation over a shaft produce a draught which the incoming air rushes to meet, and thus they powerfully promote ventilation.

The old-mining engineers of the North, provincially termed *Viewers*, were wedded to the furnace ; but as they have died or retired, their successors are more open to comparison and to conviction. Some fans have been tried and approved in Welsh collieries, and a committee of the Northern mining engineers have been engaged in instituting an enquiry, and a comparison between furnaces and fans. This enquiry ought to have been instituted some years ago, and we ought now to know whether fans really should supplant furnaces.

The gross abuses which had previously existed in many of the coal-mining districts by making one shaft do the duty of two, thus affording an insufficient supply of air, and at the same time incurring serious and constant risk of derangement or explosion, have in part been remedied by an Act of Parliament (25 and 26 Victoria, cap. 79), which rendered it illegal, after 1st January, 1865, to work any coal-pit wherein more than twenty persons were employed at one time with less than two shafts or outlets, by which distinct means of ingress and egress should be available to the persons in the pit. All our readers will remember how, for want of such distinct means of egress, the unhappy miners engaged in the Hartley pit in Northumberland were caught as in a trap by the breaking of a huge iron beam in the only shaft.

Two shafts, however, are but an elementary and a primary provision for safety, or for escape when an accident has occurred.[1] While systematic and admirably

---

[1] A subject imperatively demanding attention, is the propriety of limiting by legislature the area of coal to be worked by two or more shafts, and of restricting the number of miners employed for each pair of pits. ‘For many years,’ says Mr. Inspector Brough, ‘my own opinion has entirely

contrived ventilation is obtained in most of the well-con-
ducted pits on the Tyne and the Weir, and in some other
districts where capital and enterprise are abundant, the
reverse is still the case in several of the coal-fields more
remote from great cities and centres of skill and activity.
In passing from the consideration of ventilation it may be
remarked, that this is the principal provision on which
all practical miners rely for the safety and salubrity of
coal-pits. All others are subordinate or subsidiary to it.
By the curious and effective method of 'splitting' main
currents of air, and by the creation of various kinds of
stoppings, doors, arches and 'brattices,' a current of air
may be as easily subdivided into lesser currents as water
into separate rills; and if the position of all the pitmen
be known and 'splits' of air be taken in to them all, while
the whole of the subdivided currents are finally made to
reunite in passing out of the pit, little more is necessary
than the keeping up of such ventilation to the highest
amount of efficiency.

The question, however, necessarily arises, whether
existing arrangements can and do maintain the maximum
of desired efficiency. In all mines of whatever extent, if
not of an extremely fiery character, this maximum can be
obtained. As a general rule, it may be admitted that
thorough and systematic ventilation will accomplish all
that can be at present expected. The only serious difficulty in most cases would be the enforcing of the requisite
supervision, and the application of such regulations as
might mainly help to enforce it. But it still remains an

rested on the principle of a greatly increased number of shafts in fiery
mines.' Mr. Inspector Wynne, in a recent letter to the Secretary of State,
expresses a decided opinion that 'not more than 200 acres of coal should be
worked from one pair of shafts, and that an additional shaft should be
provided for every additional 100 acres.' It is lamentable to see the reckless and dangerous rapidity of extracting coal in many mines without
adequate provisions for safety and health.

unsettled problem, whether those pits which are extremely
fiery can, by any amount of ventilation now producible
or produced in British mines, be rendered absolutely safe.
. If we believe, with one of the mining engineers of the
North of England (Mr. T. J. Taylor, now deceased),
who investigated this subject more carefully than his
predecessors, that the fire-damp lies imprisoned in the coal
under a pressure of several atmospheres, and that it sometimes rushes out with prodigious force and volume, so as
to produce what pitmen term *blowers*; and if we further
believe that these blowers will foul a considerable area in
ten or fifteen minutes, and so foul it that the whole space
becomes to the highest degree explosive, then all such
ventilation as we have now at command must on such
occasions fail for a time, though it need not fail for a
longer time than the dissipation of the blower demands.
We agree with the late Mr. Taylor on this subject; nevertheless, we think that great blowers are uncommon; and
that by thorough ventilation, even as at present practised,
they are capable of being reduced and dissipated without
insuperable difficulty.[1] Viewers do indeed sometimes
assert that no power or skill of man can wholly prevent
explosions, and looking at the progressive deepening and
widening of dangerous mines, and the lack of correspondingly progressive precautions and vigilance, explosions

[1] When, however, we have to consider the 'goaves,' or waste and abandoned parts of old fiery mines, there can be no doubt that they are vast
natural gasometers, from the edges of which fire-damp flows out largely and
frequently. Whether a viewer should ventilate goaves, or insulate them and
shut them up, is a technical question; but those old goaves, increasing and
enlarging as they do by the mode of working pits, are most dangerous
neighbours, and are to coal-miners what powder-mills and magazines are
to country towns. Messrs. Lyell and Faraday paid a brief visit to the
North of England several years ago, and published their views on the
method of depleting goaves and rendering them less dangerous. We never
heard, however, that their suggestions were in any one instance adopted,
nor have the viewers of the North been at all influenced by their advice.

may unhappily be expected; but we have no doubt that they could be materially diminished in number, as they have already been to some extent, and that the minimum attributable to a natural necessity might be attained at no distant date.

If air be as essential to miners below as it is to us all above ground, equally so is Light. Where there is no fire-damp, the only question is, what will afford the brightest and cheapest illumination, and candles or oil-lamps may be employed at pleasure. Where fire-damp exists men can only work by the light of some kind of safety-lamp. The Davy-lamp is universally known by name, though comparatively few have seen one, or understand its principle, much less its varieties, or the improvements made upon it in safety-lamps bearing other inventors' names. The original Davy-lamp is simply an oil-lamp surrounded by a fine wire-gauze, so fine as to contain from six to eight hundred openings in a square inch. Gauze thus minutely meshed will not allow the flame from the oil wick to pass through it, and it therefore prevents the explosion of external mixtures of air and fire-damp. But it will permit such external mixtures to pass into the flame, and then there will appear a *cap* or *top* on the interior flame greater and greater in proportion to the explosiveness of the exterior mixture. Thus we may have a half-inch or one-inch top on the interior flame, and learn thereby the certainty as well as the degree of danger. Pitmen are too familiar with such tops to be greatly alarmed at them; but the casual visitor to a pit, however much he may confide in the Davy-lamp, does not feel at all easy under such appearances, and is soon disposed to return towards daylight and pure air.

The Davy-lamp is the best friend of pitmen, but they often treat it as disrespectfully and injuriously as they would their worst enemy. They attempt to enlarge the flame and to get more light from the lamp than it is formed to give. Some will even make a match-box of it, and open it to light their pipes. To preserve it from such indignities most employers put a lock upon it, and all lamps are or should be locked before delivery to the miners. Yet not a few of these men are so reckless and perverse as to procure false keys or to pick the lock of their lamp. Disastrous explosions have thus been caused, including, to all appearance, one of the recent great explosions in Yorkshire. In such instances the fool-hardy miner usually meets the death he has brought upon his innocent but destroyed companions. Let the men be as strongly reprobated for such murderous mis-deeds as their masters for other carelessness. A picklock lamp-bearing pitman in a gaseous mine is a criminal of unmitigated baseness, and ought to be punished pro-portionably to the calamity contingent upon his mis-conduct.

For more than forty years has the Davy-lamp been in use, and therefore upon public and careful trial. Gene-rally, the Northern mining engineers regard it as really safe, and we have never found experienced miners declare it to be otherwise, under ordinary conditions, and with the use of the essential precautions.[1] Nevertheless, some modified lamps are rather preferable in practice. Thus George Stephenson's lamp, which the pitmen fondly call 'the Geordie,' is protected by a glass, and so long as this exterior glass remains whole, gas will not increase the in-tensity of the flame beyond a certain degree, when it will

[1] The long and animated controversy about the safety of the simple Davy-lamp has not materially affected the opinion above expressed. Never-theless, certain particular instances of explosion do seem to tend in some degree to invalidate the prevalent confidence; but not unless there are strong currents or extraordinary conditions. On the whole, then, this lamp may still be pronounced safe under ordinary conditions; and Sir Humphry Davy himself does not appear to have expected more.

be extinguished. Should the glass break, this lamp simply becomes a Davy-lamp. The glass may be broken accidentally, but in a mine where 200 such lamps were in use, it has been found that scarcely ever has a piece been broken out of a glass.

Another form of lamp is that invented by Dr. Clanny (whose claims were rather slighted); and when a metal chimney is given to Clanny's lamp, with some trifling additions, it becomes the lamp of Mueseler, which is much in favour and use in Belgium, where 400 are employed in one mine. Still further modifications in safety-lamps have been introduced both at home and abroad. The grand desideratum is to obtain more light consistently with security. The objection of the pitmen to the Davy is lack of light, but light must always be combined with security and portability. In respect to safety, undoubtedly those lamps which go out at once in an explosive mixture are the best; but safety at the cost of being left in total darkness in the inextricable labyrinth of passages in an extensive mine, is extremely inconvenient.

It is manifestly bootless to put confidence in safety-lamps while *gunpowder* continues to be used for blasting the coal in fiery pits. It is inconceivable that so perilous a custom anywhere prevails. Yet even so lately as 1862 sixty lives were sacrificed by it at St. Edmund's Main Colliery near Barnsley in Yorkshire, and other serious accidents have probably been due to the same cause. Such an infatuated proceeding is actually conveying flame to the gas, and doing on the one hand what, on the other hand, safety-lamps are designed to prevent. The 'firing of a shot' in the inner working faces of a pit known to harbour much fire-damp is no ordinary trial of courage. The sound of the dull boom that follows each blasting conveys to all who are not habituated to it the idea of a more extended and fatal explosion—that of fire-damp. It may be thought that none but madmen would thus blast coal while any quantity of gas is or has lately been present; equally, however, may it be thought that no sane man would, in like circumstances, open his Davy-lamp; yet men do both the one and the other. As it is affirmed that fire-damp sometimes issues very suddenly in large quantities, and fouls a whole district of a pit in perhaps ten or twenty minutes, getting coal by gunpowder should be absolutely prohibited whenever gas comes forth so abundantly as to require the use of safety-lamps.[1]

One of the foremost among the many advantages of the Davy-lamp is its value as a sure and simple indicator of the presence of fire-damp, and therefore of danger. Every morning, before work begins, a subordinate officer, named an *overman*, or *deputy*, perambulates the working places of the pit, and, with his Davy in his hand, carries the test of safety and the measurer as well as the indicator of danger. If his lamp show a tall cap or a halo and an enlarging flame, he at once knows that the hewers of coal who are to follow him must be warned of peril, and in case of much explosive air a rude signal is so placed as to warn off the hewers altogether from the place. Men who disregard this warning do so in defiance of death. The overman himself incurs the first risk, and measures it, and hence he ought to be much better

[1] In their joint report, dated 26th of Jan. 1867, the united inspectors recommend that this new General Rule be added: 'In all workings in coal, where safety-lamps are used as the means of lighting, no blasting powder shall be used in such mine.' We should also be inclined, with one of the inspectors, Mr. Brough, to introduce a rule against the mixed use of safety-lamps and naked lights in gaseous mines. The use of candles, either by exploratory or other miners in pits at all fiery, leads to numerous fatal accidents, all of which might be prevented by the exclusion of naked lights. A General Rule to this effect appears to be advisable.

acquainted with the principles of the Davy and the chemistry of gases than he commonly is. Upon his preliminary survey, and that of his colleagues, depends the safety of the whole company of workmen underground. Many of these deputies are uneducated and ignorant to a lamentable degree.

Yet even as an indicator the Davy lamp has its limits of usefulness, for it can only indicate to the one bearer of it, if he be an overman alone on his early travels through the pit, and it can only indicate at one time as well as to one man. To become a continual indicator, it must be suspended *in situ*, and frequently consulted. To compensate for these defects, which necessarily attach to all portable lamps, Mr. George F. Ansell, of Her Majesty's Mint, has invented an ingenious and simple instrument, by which he professes to show both the accumulation and the rate of accumulation of a mass of fire-damp, or even of carbonic acid, so as to make this known to the master without man's agency, and thus to cause the servants to be more particular in observing rules and performing their duty. For this purpose Mr. Ansell employs a balloon of thin india-rubber, which, singularly enough, is expanded both by carbonic acid and by fire-damp. This balloon is filled with atmospheric air, tied at the neck, and bound round to prevent lateral expansion. So prepared it is placed under a small lever upon a wooden stand, so as to exert a gentle pressure on the lever. If fire-damp or carbonic acid accumulate round it, either the one or the other will pass (by a particular law of gases well known to chemists) through the substance of the india-rubber balloon, and, accumulating inside, cause it to expand, so as to press against the lever, thereby raising it and releasing a detent, with which the terminal poles of a battery are connected. By this arrangement a telegraphic communication is effected to a distant locality, and a warning on the spot may be given simultaneously

The greatest peril in a fiery mine, and that which is the most difficult to be guarded against, is a sudden and violent discharge of fire-damp. Such discharges, as we have previously intimated, are known as ' blowers;' they issue with prodigious force and emit an immense quantity of fire-damp in a very short time, so that long passages of a pit have in not many minutes become filled with highly explosive air, or, as pitmen say, ' fouled.' In such cases Mr. Ansell would apply a little apparatus in the belief that no eruption of gas can be so sudden that his invention will not announce its approach in from five to ten seconds, according to the percentage of fire-damp contained in the eruption. This instrument consists of an iron funnel, whose stem is bent like the letter U, the funnel part being closed with a plate of glazed Wedgwood ware; the stem being closed with a cap of brass, through which is passed a platinum-tipped copper wire, capable of just dipping into mercury previously placed in the bend of the funnel. The distance between the platinum-pointed wire and the mercury regulates the point at which the indication as regards the eruption should be given. If, when the instrument is properly adjusted, gas impinges on the porous tile, diffusion (according to the known chemical law of diffusion of gases) takes place, and the presence of the accumulating gas forces the mercury against the platinum-pointed wire, and the circuit being thus completed telegraphic warning is given on the spot, and may be conveyed to the distant manager's room, either by a needle or by a bell.[1]

Such instruments are admirable in theory, and ingenious in their useful application of the remarkable law of the diffusion of gases, made known by Mr. Graham, but it remains to be seen how far the promises of their inventor are verified by practice. At least full and repeated

[1] Appendix No. 4 to the Evidence of the Select Committee on Mines.

trials should be given to them before they are consigned
to neglect.   Some similar or modified instrument may be
ultimately found successful, and it would be no slight
advantage in the management of a fiery pit, if the
manager could be infallibly apprised of accumulating gas
even though its amount were unknown.   A further and
signal benefit would be conferred by enabling the
manager also to ascertain with facility the amount of gas
per cent., and this Mr. Ansell would accomplish by employing the present aneroid barometer, very delicately
constructed, and bearing a porous tile in the place of its
ordinary back.   Having adapted it for his purpose by
some mechanical additions, he takes it into the suspected
part of a pit, and holds it by the ring handle till it has
become of the same temperature as the place where it
then is.   The maximum effect of a gaseous atmosphere
on this instrument is produced in about forty-five seconds,
when the face of the indicator must be immediately and
accurately read, because when the maximum is attained,
what the chemist terms *effusion* takes place, namely, the
mechanical passage of gas through the tile by pressure
above, and then the hand on the face travels back to
zero.   When *diffusion* ceases, effusion takes effect and
empties the chamber of gas.

Our apprehension is, that in the present lack of scientific knowledge, accuracy, and tact among at least the
second and inferior orders of colliery managers and men,
Mr. Ansell's instruments are far too delicate in their construction and application to become much used in pits;
but he deserves great credit for his philanthropic attempts,
and practical chemists will admire his ingenuity if practical colliers decline to avail themselves of it.   It is
notable that the Davy-lamp simultaneously afforded corresponding indications while Mr. Ansell was testing his
aneroid instrument, and it is satisfactory to find, as Mr.

Ansell observes, ' that marvellous arrangement, the Davy-lamp, gives magnificent indications.' In truth, if the simple Davy-lamp were fairly treated and fully accepted, and if those who use it were thoroughly acquainted with its principles and with all its indications, fiery pits might be worked with comparatively little risk of explosions.

An electric light has naturally suggested itself to men of science as a source of illumination for coal-pits, and two or three years ago a foreign *savant* appeared at Newcastle with a mining lamp on this principle. This invention gave a safe light, but too feeble and uncertain for practical use; it had been suggested to the inventor by the glowworm. On the general question there can be no doubt that a powerful, brilliant, and perfectly harmless light might be introduced into pits by the use of a good battery and all its necessary adjuncts. But no man who has travelled through the long, weary, tortuous, and seemingly interminable ways and bye-ways of an old and extensive coal-pit would propose batteries and wires and other apparatus as a *simple* substitute for portable safety-lamps. The first question would relate to practicability, which may be doubtful from the extent of area, the difficulty of adjustment, and liability to derangement. The next question would be cost, and on that we have no doubt that the expense would deter nearly all managers and owners of coal property from voluntarily employing the apparatus. The third question would be, Who will consent to bear the expense and pains of a first and full trial? It would, however, be very desirable to make experiments on a small scale, and thus to ascertain if these warrant a larger outlay and more extended trials. Pitmen have a great objection to any but the simplest and most indispensable apparatus underground, and numerous or lengthy wires would be much exposed to casual or malicious injury.

A public exhibition of mining implements and improvements in Newcastle or Durham or Leeds, Manchester, or
Birmingham, would be of great utility to the coal-mining
population.   At such an exhibition managers and miners
might observe models or specimens of every kind of
safety-lamp, and of all implements and instruments which
have been suggested either by scientific or practical men.
If such exhibitions were periodical, and if foreign coal
countries were invited to combine, signal advantages
would ultimately accrue.   It is strange that so vast and
important a section of our industrial population as our
coal-miners—the males, according to the census of 1861,
employed in coal-mines, amounting in number to 282,473 ;
and according to the computation of the inspectors in
1865, to 315,451—should be without some such exhibition, while agricultural shows, displaying various agricultural implements, are frequent and successful.   The
numerous benefits arising from the adoption of this suggestion will occur to miners, and one at least would be
the establishment of intercommunion between the men of
different and distant coal-fields.   Thus some of the
33,000 colliery workers of South Durham might frater-
nise for the best purposes with some of the thousands of
their fellow-workpeople in other localities.   Scientific
men, too, might attend these meetings, and there might as
well be a British Association for the Advancement of
Mining, as a British Association for the Advancement of
Science.   All projects like those of Mr. Ansell might be
there expounded and discussed ; and instead of mere
local Institutes of engineers there would be a general
gathering of mining science and skill from our entire
coal-fields.   There was, indeed, a few years ago, a meeting of Mechanical Engineers at Birmingham, which was
both successful and agreeable.   Why should there not be

an extension and recurrence of such meetings, on the plan of our suggestions?

While so much has been said of fire-damp, it is apt to be forgotten, and perhaps is little known, that mining viewers of large experience attribute at least seventy per cent. of the deaths in fiery mines to *after-damp*, and the small remainder only to fire-damp. The after-damp, sometimes called choke-damp, is a combination of gases in proportions dependent upon the peculiar circumstances of each explosion. These combined gases are nitrogen, carbonic acid, carburetted hydrogen, and common atmospheric air. Carbonic acid gas is dense and poisonous, and in an atmosphere containing as little as ten per cent. of it, human life can only be maintained for a short time. Indeed three per cent. is fatal if the amount of oxygen falls below eighteen per cent. in the same atmosphere.

In several instances of exploded pits, little companies of men in the interior who had escaped and survived the explosion have been cut off and destroyed by·the immediate generation of after-damp. The effect of death by the one gas or the other is very distinctly seen in the countenances of the dead. The poor men killed by the firedamp are marked with burns and scorchings, and their features all more or less distorted or disfigured. On the other hand, where men have been suffocated by chokedamp, their features are placid and simply inanimate. It is evident that the after-damp has speedily deprived them of the power of breathing, and has almost instantly choked them.

Utter neglect of precautions against after-damp seems to be prevalent in all mining districts. Not a few propositions have been made from time to time in this direction, but we never read of any practical application of them in the hour of danger and death. If common air could have been conveyed at once to the two hundred

and four sufferers in the Hartley pit when they were
under the deadly influence of after-damp, most of them
might still be alive.

It should be the aim of managers by additional con-
trivances to render every miner as independent of the
noxious influence of after-damp, as a good safety-lamp
can make him independent of fire-damp. The miners
themselves should always carry simple means of covering
the face. An inspector of mines once told us that in
investigating the circumstances of many explosions, he
heard of several instances where pitmen caught by an
explosion, but surviving it, had placed a wetted cap or
handkerchief over their faces, and then passing over the
dead bodies of their unfortunate companions, had come
out safely from the extreme recesses of the pit. So, too,
have pitmen saved themselves by stuffing their caps into
their mouths, and then bending as low as possible have
crept securely to the shaft.

Those to whom this subject is unfamiliar will have
been astonished to perceive in the brief tabular abstract
at the commencement of this article that the deaths from
falls of the roof and pieces of coal in pits have amounted
to 3,953 in ten years, and that these form two-fifths of
the entire number of deaths from all extraordinary
causes, and considerably more than those arising from
any other specific cause, being at the rate of 39·87 per
cent. The difference between the mortality from explosions and from such falls is similar to that between a
brief epidemic and an habitual local disease, or, to vary
the metaphor, between a short, sudden battle, and a continuous dropping fire upon a retreating enemy. A little
explanation will make it apparent how these deaths so
frequently, and, as it may be said, so regularly, occur.

Nature has packed up coal so closely and compactly
that it is impossible to put more of it in the same space

by any mechanical means. But when man begins to
break up a coal-seam, he, as it were, unpacks the load, or
destroys the coherence and balance of the whole mass.
He removes the natural support of the coal-bed; he
takes, as an architect would say, the floor out from the
building, and as he only imperfectly props up the roof
and walls, a more or less speedy ruin ensues.   The main-
ways or high-roads of a mine are built up and arched like
our railway tunnels, and hence they are safe enough.
But this cannot be done in the low, long, narrow passages,
which are often not above four feet in height, nor could
proprietors bear the expense for many miles in area.
Propping up with timber is the only available expedient,
and when sufficient and sufficiently sound timber is
employed little more is needed than watchfulness in
repairs.   Unhappily the temptation to spare expense
often leads to inadequate timbering; and the rapid destruction of the props by wear and tear, and the rot
generated by the damp heat underground, make this a
serious item of expenditure.

The character of the coal itself in some places adds to
the necessity of much timbering.   In South Staffordshire
the rhomboidal structure of the material in its jointy
beds causes the famous ' ten-yard seam ' to be much
intersected, and produces a naturally complicated network
of joints, in the direction of which the coal is very ready
to fall asunder.   This corresponds in some measure to what
the Germans expressively call ' slickensides.'   Moreover,
by what is technically termed ' the pillar and stall ' mode
of excavating, enormous cavities are left behind, often of
thirty feet in height, and of course ' settlements ' ensue,
together with violent concussions and sudden collapses,
locally called ' lumps,' of great masses of material.
Even a small piece falling from a height of from twenty
to thirty feet may kill a man. How much more fatal must

large fragments be! In the year 1857, in the collieries only of South Staffordshire and Worcestershire, eighty-one persons were killed by falls from the roofs of pits, in obtaining 6,000,000 tons of coal. This amounts to 13·5 deaths per million ton of coals got. At the same time, the average of deaths from the same cause for the whole of Great Britain was 5·15. Thus, this is the special evil of one district, and the local inspector has stated that inexcusable carelessness prevails in timbering the roofs. If a place only looks safe, without any test, a few props alone are put up, and consequently a fall soon occurs. 'The reigning cause of the destruction of human life,' to quote Mr. Brough, an inspector's words, ' is the constantly recurring falls of stone and coal from overhead and from the sides of working places.' [1] Indeed, if we take England, Wales, and Scotland together, it may be affirmed that not a single day passes without the occurrence of some calamity from falls of ponderous masses of coal or shale or ironstone in coal and ironstone mining.

No legislative interference with the systems of excavation would be tolerated,[2] although it is affirmed that the ' long wall' plan is superior to the ' pillar and stall ' mode of working, in particular cases at least, for by the former more coal is raised per acre, and better ventilation can be secured; but adequate timbering does come within

---

[1] The same inspector thus reports, under date February 1, 1867 :— 'Something really must be done to prevent mortality by these falls of coal and stone. The number annually killed by such class of accident is dread-ful in the extreme. Taking an average number of years since the inspection was established, it will be seen that death by falls of material goes fright-fully beyond any loss of life by explosion of fire-damp. It is clear that something must be done: we should not go on crushing the people to death in this way.'

[2] In their Petition the miners complain that 'the practice in the Stafford-shire collieries of working the thick coal in more than one face is highly dangerous and very destructive to life in the said collieries.'

the province of Government inspection, on the ground of safety of life and limb, and it is a miserable parsimony that permits one life to be destroyed every day in the year by culpable neglect.

There is a very dangerous duty in the economy of timber, called 'drawing the props.' In so doing men stand ready with an axe in the waste places of a pit about to be abandoned, and knock out each prop successively, rescuing it and themselves at the same time. In one pit we ourselves stood at a little distance and witnessed this process. A man cleverly struck a prop, while another drew it away, and both retreated. The roof then, having nothing to support it, began to give way and fall in. Woe to the unfortunate prop-drawer who cannot escape before the stones fall; yet men can be got to perform this hazardous duty for about the same pay as the hewers of the coal obtain.

The day, though not the date, may be clearly foreseen when machinery for cutting coal will supplant, so far as hewing is concerned, human labour. Coal-cutting machines have been for several years proposed and displayed in one or two forms, and have been successfully tried in a few instances. But the spur of necessity and the expectation of independence will urge on the masters to quicken and foster the germs of such inventions. At this very time the South Lancashire and Cheshire Association has offered prizes of money for the best machines of this character. All the efforts of the working miners to act upon and control their employers will tend to this end, and slow as the masters may be in entertaining other projects, they will gladly welcome such machines as will render them more and more independent of their men. Thus inferior and ungenerous motives often lead to substantial improvements.

Many arrangements have already been made underground, by which human suffering and painful labour are diminished. We remember a collier's song which had the burden, 'God bless the man who brought us metal trams,'—that is, the tramway-plates of metal which of late years have been laid down in pits, and have enabled the 'putters' to push along their loaded baskets of coal with greater facility. Much more animated and choral will the burden some day be, 'God bless the man who brought us down machines.' Meanwhile, however, adjustments consequent upon the introduction of new machinery will have to be made, and temporary deprivations to be endured. But hewing down coal in deep, hot, dark, dusty corners of pits is not a kind of labour which we can honestly wish to be perpetuated for civilised men. It is not, perhaps, so unhealthy as some other occupations, chiefly because it is carried on only for a few hours per diem consecutively. But in thin low seams, where the hewers must needs lie on their sides, or squat down half naked in painful and unnatural postures to perform their work, the sooner cutting machinery can be introduced the better—better for the men as well as the masters, better for the mind as well as the body.

That the number of fatal accidents in shafts of coalpits should amount to 1,710 in ten years—that is, to nearly one-fifth of the total number of accidents in the same period—may strike many readers with surprise. Had they, however, themselves descended shafts, especially in ill-conducted mines, that surprise would be considerably diminished. The passage below and upwards in many inferior mines is a passage of perpetual perils. The chain, or the conveying vehicle, or the rope, or some part of the winding apparatus, may be suddenly broken, and death as suddenly ensue by a terrible fall of the living load. From the careless way in which some men and many boys ascend and descend they are in daily risk

of such a calamity. Even without carelessness in themselves, it may exist in the man who superintends their
conveyance, or some part of the gear of the engine or
pumps may fail. What happened in 1862 in the shaft
at Hartley is too clearly in remembrance to need more
than an allusion. A great variety of accidents may and
do happen in shafts, and many that are injurious though
not followed by death.

Nearly the whole of these calamities, and perhaps
really all of them, arise from preventible causes. There
can be no unalterable reason why any human being should
fall down a shaft, or why a stone from the shaft-side, or
the passing load of coals, should fall upon him. Ropes,
chains, buckets, and all winding gear, can be made strong
and good, can be renewed when old, and examined when
doubtful. The improvements, too, in shaft apparatus are
considerable. Strong wire ropes are now and long have
been in extensive use, while buckets and crazy baskets
are superseded by the so-called 'safety cages,' which run
smoothly upon iron guides, and somewhat resemble
vertical railway trains. While descending in these safety
cages one feels in perfect security. True, the suspending
chain may break, but the cage is instantly caught by extending arms and stopped. All such machinery being in
the first instance costly, it is chiefly adopted in superior
districts, and is seldom found in rougher and more primitive coal-pits. In these latter, moreover, there is a lack
of sufficient fencing and guarding, and consequently there
are a number of what are in effect pit-traps set for the
destruction of any luckless and unwary passenger. The
Act provides that none except males above fifteen years
of age shall have charge of a windlass or other winding
gear, and that in certain cases steam engines for winding
shall only be in charge of persons above eighteen years
of age. So far good, for mere lads were previously in

charge of such apparatus; but it is to be hoped that fatal
shaft accidents will now more speedily and very considerably diminish.

What are termed 'Miscellaneous accidents in mines
and above ground,' which amounted to 2,234 in the ten
given years, and formed more than one-fifth of the whole
number of fatal casualties in that time, result from a
great variety of causes. These we cannot specify in
detail, but they are all, like those previously treated of,
preventible by vigilance, care, and forethought; and even
if this assertion were questioned or qualified, there is no
doubt that a very important diminution might and should
take place in their number. We do not mean to assert
that with the present workmen and pitmen all recklessness
or inattention can be speedily eliminated. But the more
these men are educated and raised in self-respect and
sensibility to evil, the more will the number and the
occasions of injuries and fatalities be lessened.

From the information afforded in the preceding pages
the reader will be qualified to consider and form an
opinion upon the important subject of coal-mine inspection. It was a matter of great difficulty to obtain
even a qualified assent from the coal owners and managers
to any amount of inspection, and incoherent as these
authorities were on other subjects, they became coherent
enough for opposition to what they regarded an inva-
sion of private rights and liberties. A landowner who
managed or possessed a coal mine treated it like a game-
preserve, and warned off intruders by penalties and
threats of man-traps and spring-guns. The first Children's
Employment Sub-Commissioner who visited the great
northern coal-pits, found himself suspected, misconceived,
and regarded in the light of a superior detective. More
light gradually broke in upon the minds of the mining
magnates, who became more urbane; but so soon as the

first Act for Inspection was about to be framed, the coal-lords and the principal northern viewers combined to offer a strong opposition on details. Even enlightened Nicholas Wood took up arms in this warfare, and although it was never known to the public, every important clause in the Act was a matter of contest and mutual concession. The Act therefore finally resembled a treaty between belligerents, and though what is now secured is less than one side hoped for, it is more than the other was at one time inclined to yield.

Now, however, that several years have elapsed, and inspection has been found to work well on the whole, we are in a much better position to judge of its efficacy and of its possible extension. In the tabular statement of colliery accidents for ten years before alluded to, which Mr. Atkinson, an Inspector of Mines, presented to the Select Committee on Mines, there is a comparison of the results of two quinquennial periods, and this comparison is designed to display the benefits of inspection. During the first period of five years embraced by the table, from 1856 to 1860, both inclusive, there were 5,089 deaths in the kingdom from colliery accidents; and during the same period 381,067,047 tons of coal were raised. On the other hand, during the next five years, from 1861 to 1865, both inclusive, 468,548,905 tons of coal were raised, so that if the deaths had increased in the same proportion as the quantity of coal raised, the deaths during the latter period of five years would have been 6,257·3 in number; whereas they only amounted to 4,827, being 1,430·3 fewer deaths during the second than during the first quinquennial period. This is a reduction to the extent of 22·9 per cent. in five years, being at the rate of 4·58 per cent. per annum in relation to the coal raised.

To display the benefit of one measure, viz., the passing

of the Duplicate Shaft Act in 1862, it is shown that during a period of three years (1860, 1861, and 1862) immediately preceding the operation of this Act, there were 3,178 deaths from colliery accidents throughout the kingdom, while during the same period 264,358,164 tons of coal were raised; whereas during the three years (1863, 1864, and 1865) succeeding the passing of the Act, 286,853,443 tons of coal were raised; so that if the number of deaths had increased in the same ratio as the coal raised, the number of deaths during the latter triennial period would have been 3,448·4, while the actual number was 2,758 deaths, being 690·4 deaths less during the second than during the first of these two consecutive triennial periods.

Statistics and verbal testimony alike demonstrate the value of inspection. In some districts the value is more conspicuous than in others, and it may not be wholly attributable to the actual work of the inspectors, but as much to the greater care induced in the conductors of collieries, by the knowledge of the law as well as the beneficial operation of its provisions. For example, there are general rules laid down by the Secretary of State, and in addition special rules for each district administered by the authority of the inspector. The special rules for the Lancashire coal-field, now lying before us, are well ordered and carefully detailed. By general and special rules, therefore, there is now a sort of codification of colliery law. The general rules for all coal and ironstone mines are fifteen in number, and are printed in the Act. The special rules are in each case to be framed by the owner of every coal and every ironstone mine, and transmitted to the Secretary of State for the Home Department, after having been publicly suspended for fourteen days; and are finally returned to the framer for constant exhibition and enforcement. All the provisions relating to rules are excellent.

At present there are twelve inspectors of coal mines for England, Wales, and Scotland, and it has been the aim to distribute their allotted duties as fairly as could at first be estimated. In practice, however, great inequalities are found to arise, and must from year to year arise in consequence of the difference of districts and the disproportionate scales of mining extension. An inspector who has a number of fiery pits in his province experiences vastly more care and anxiety than one who has none of that class. He is always at the mercy of fire-damp, and always in fear of an explosion. Such an explosion as that which lately happened at Barnsley awakens grave solicitude and occasions much extra work for an inspector. In this particular instance the inspector resigned after the investigation into the explosion, and has been succeeded by another officer.

Each and all of the inspectors have far too many collieries to inspect if they really attempted to examine them. No doubt the public, when they think for a moment on the subject, are under the impression that every considerable coal mine in the country is in course of periodical examination by some one of these twelve official gentlemen, and no doubt this is a natural conclusion. The inspectors themselves, however, hold a very different opinion of their duties, and carry out a very different practice. They do not profess to descend into and thoroughly inspect all of their pits. [1] They do

[1] Although the above observations are still in a measure applicable, yet by the Coal Mines Regulation Act of 1872 (35 & 36 Vict.), the Home Secretary is empowered to appoint assistant-inspectors, and has already appointed several who undergo a previous examination as to competency. It is intended that these gentlemen shall work under the direction of the chief inspectors, and perhaps ultimately succeed them. So far the suggestions offered in this article have been accepted.

not even propose to go down and through them all in any stipulated period. Many of the colliers have never seen them at all, and do not know them. To 'visit all the places underground is an utter impossibility as things now are, and an inspector may live and do his work during ten or fifteen years in his locality, and make his annual report to the satisfaction of the Home Secretary, without a personal and intimate knowledge of more than a very limited number of his collieries. All this they themselves admit, and this is what the working miners, as most people will think justly, complain of.

Can and ought 'the present twelve inspectors to do more? They themselves would reply—we cannot and we ought not. We cannot, as a matter of physical capability and personal comfort—we ought not, as we think, on grounds of expediency. We cannot, because we are not ubiquitous—and because it would be manifestly impracticable for us to satisfy ourselves of the condition of every mine out of some two or three or four hundred under our charge. Besides, they would add, it must be inexpedient for Government, through us, to assume the direction or admonition or control of all the mining managers subject to our visits. In such cases we should become the head engineers of our districts, and have to come into frequent collisions with the several managers and owners, with whom it is our policy to keep, as far as possible, on good terms. If we were to attempt to perambulate the whole of the deeper and larger mines, each of them would demand a day, or a large portion of a day, and we could not accomplish our other duties. One of these gentlemen, indeed, openly declares that if a condition of inspection be that the inspector may be called from his bed at dawn, to go down mines all day, the sooner the Act is repealed the better for all parties. A perpetual peripatetic supervision is to be dreaded alike by those who would have to perform it, and those who would have to endure it.

Here comes in the consideration of the objection founded upon a diminished or shifted responsibility. If, say the masters and managers of collieries, you propose to establish over us a rigorous and minute inspection, extending to the enforcement of the inspector's opinion or advice in details of management, then you will shift the responsibility of safe working from us to the inspectors. We cannot follow their advice and at the same time be responsible for the consequences of doing so—especially if their views should be contrary to our own. We, for instance, may think that a certain course of proceeding recommended by an inspector would lead to danger or an explosion. If we are compelled to adopt it, and an explosion should ensue, clearly we should be guiltless. We are satisfied to have the near residence and occasional visits of an inspector, and we do not object to show him whatever he legitimately desires to examine. We want no more, nor should we willingly endure more.

Thus both managers and inspectors come nearly to the same conclusion. Things are tolerable as they are, are better than they were, and, moreover, may be still better in years to come; why then interfere and introduce a new element of discord?

On the other hand, the miners themselves, as represented by the present petitioners and the most respectable and reflecting individuals of their body, express or would express, different views. The actual terms of the petitioners are these:—

That the fearful sacrifice of life in mines and collieries affords abundant proof that the legislative measures hitherto passed have proved totally inadequate for securing the personal safety of the miners of this country.

That your petitioners believe that the sacrifice of life in mines

and collieries can only be prevented by the appointment of a
sufficient body of sub-inspectors, whose duty should be to examine
internally the mines and collieries in which your petitioners are
called upon to labour.

That the constitution and practice of coroner's courts, so far as
they relate to accidents in mines, are so objectionable that justice
is not secured to your petitioners.

That accidents in mines are mainly caused by the want of skill
in the agents, overmen, and chief managers of mines and collieries,
and from lack of diligence or want of care on the part of the subordinate officers.

Such are the words of the ' under-miners '—that is, we
presume, the underground miners—of Northumberland
and Durham, who as a body are considerably in advance
of similar men in other coal-fields, and who, it should be
remembered, work in the best regulated mines, on the
whole, in the kingdom. The evils, therefore, which exist
amongst them may be fairly presumed to exist in a more
aggravated degree in other collieries.

Whatever abatement we may make in the representations of these miners for their supposed *esprit de corps,*
and for their presumed antagonism to their employers—
which, however, we are not inclined (from personal visits
paid to these pits and pitmen in the two counties named),
to estimate as very important in this particular cause—
whatever abatement may be made for any such feelings,
there remains a very large residue of fact which the
miners affirm they can justify by their experience. When
examined as witnesses they naturally appear deficient in
tact, and in the power of stating their case strongly. So
that we must not judge by the scanty evidence we get
from them in the volume before us. Knowing, however,
full well their views and their mode of life and labour, we
may state their case, perhaps, with more force and clear-
ness than they do for themselves—and at the same time
concisely.

They feel that the present supervision is quite inadequate for their safety, because they know from daily experience what is deficient; they work in the midst of such
defects, and hear each other's complaints. The miners
know what the inspectors do not and cannot know, except
from themselves. It is true that they can give notice to
the inspectors in their several districts of a notorious defect and any apprehended danger, and that in most
instances they can rely upon the inspector's attention to
their notification; but at the same time they feel that by
giving such notice they incur the secret or open dislike
of their superiors, and are sometimes plainly told, ' If the
pit is not good enough for you, you can seek another.'
Unfortunately they often have no personal knowledge of
the inspector, and are seldom aware when he is about to
visit their pits, although they allege that the masters
always are, and often prepare the pit accordingly. They
desire to have opportunities of acquaintance and conversation with the inspector, and such ready access to him
as shall enable them to tell him freely and instantly when
there is danger and discomfort; and they desire to be
allowed, without identification and risk of dismissal, to
point out where and when his presence is required. They
know so well, and see so clearly, that a well-designed
system of inspection may be nullified by stratagem and
avarice, that what really looks well to the public, and
works well to some extent, appears to their strong in-
sight ' totally inadequate to securing their personal
safety.'

They do not so much wish for more inspectors of the
same order as the present ones, although they would
gladly welcome them if competent, as for sub-inspectors
who would probably be more freely appointed and easily
found than principal officers. Such persons, indeed, are
ready to hand in and around the pits at this time. The

working miners would suggest the appointment of such
men as the under-viewers or second-rank managers of
pits now are. This class of men (speaking from what
we remember of the Northern collieries) obtain salaries of
from 150*l.* to 250*l.* a year at present—rarely, perhaps, so
much as 300*l.* They have in nearly all cases been brought
up to the duty from early years, and in actual pitwork
are frequently clever and trustworthy. Indeed, from
this class often and ultimately come the viewers or head
managers themselves. Select from their ranks twenty or
thirty of the best men, subject them to careful examination, throw the appointments open to competition, and
when you have obtained the required number, then distribute them over the entire kingdom and adjust their
duties in subordination to the present and future twelve
inspectors-in-chief. A systematic and personal inspection
of pits would then be in operation; the sub-inspectors
would report to their principals, and the labour of the
latter would not only be lightened, but better directed,
and not as now often dissipated and spent in vain; at the
same time the working miners would feel more at home
with the sub-inspectors, would get frequent hearings, and
fuller sympathy with the details of their complaints.

The question is not whether the masters, owners, and
chief engineers of collieries approve of this suggestion,
for they certainly do not; nor is it a question whether
there would not be frequent discords and disputes in consequence. Assuredly any regular and rigorous examination of pits would lead to these consequences upon the
discovery and indication of neglect or defect. But the
true question is, would the plan be efficient, and would
the ideal of colliery inspection be thereby carried into
practice? Readers can judge for themselves, to some
extent, from what we have laid before them; and we
certainly do not see the full force of the objections of

masters, managers, and inspectors to such a proposal.
If they could be obtained, and paid, and stationed without delay or difficulty, we should prefer an addition to
the present body of inspectors on an equal footing.
Twenty in place of twelve might accomplish much, and
in no long time bring up the inferior mines to the present
level of the superior ones. Twenty capable and earnest
men—twenty such men as *some* of the present inspectors
are—ought fully to carry out the provisions of the
existing Act—suggest and secure improvements, and,
if fully empowered, work out all that could be fairly
expected from supervision.

But if there be financial or other impediments to such
an addition to the inspectors-in-chief (who, we may note
in passing, are not too highly paid, if really efficient),
then let there be sub-inspectors. The experiment has
been tried in the inspection of factories, and has worked
thoroughly well, to the satisfaction of all concerned.
This fact is lost sight of when a number of fanciful
objections are raised to the project, and when it is pronounced inoperative. Apart from possible improvements
in inspection, and reverting to the present time, there is
manifestly a want of some considering, experimenting,
criticising, and admonishing head, or association of heads,
in relation to the safe conduct of our extensive and
valuable coal mines. There are now 3,256 collieries in
Great Britain,[1] and so rapidly have these increased in
number, that in 1865 the number was only 2,614; should
the future increase be in the same ratio, which perhaps is
not likely, the argument we are now employing would
be by so much strengthened. In all these collieries, in
effect, every owner does as seemeth to him good. He is
circumscribed only by conditions of policy, and to a

[1] In 1871 the total number of collieries under Government inspection in
Great Britain was returned as 2,760.

slight degree by the few Government officers entitled
inspectors.   True, efficient superintendence, like honesty,
is the best policy, for an owner of mines loses far more
by a serious accident than he would pay for good over-
looking.   The same observation applies to fire insurance,
yet many men do not insure; and, in like manner, many
colliery owners do not provide the most careful, and consequently the more costly, supervision.   There is wanting,
therefore, for such men a force *ab extra*, and knowledge,
judgment, and skill to justify that force.

The most delicate and successful indicator of fire-damp,
the most improved and approved safety-lamp, effective
ventilating fans, excellent and safe shaft machinery and
winding-gear, superior underground apparatus, coal-
cutting machines of approved power and economy, and
all the various coal-pit improvements and ameliorations
which have been for so many years proposed and discussed—partly adopted and partly neglected,—all these
may be well known to men of science, and their respective
merits and defects clearly ascertained.   Yet so far as
most coal owners and managers of pits in general are
concerned, they might as well never have been invented
or considered.   There might be a Pantechnicon of
coal-mine machinery and mining-gear in London, and
every implement and improvement might be registered
and ready for use, but still, without the impulse *ab
extra*, things may go on underground as negligently as
of old.

Strong hopes were once entertained that a special
Mining College would be established somewhere in the
North of England.   For years the project was discussed,
half-formed, and much favoured by the mining engineers of
Newcastle-on-Tyne, and principally and earnestly advo-
cated by the most eminent of them, the late Nicholas Wood.
Even he, however, with all his experience, influence, and

skill, failed in this excellent enterprise ; with his decease it is to be feared the Northern Mining College in prospect also perished.[1]

The apathy of large bodies of mercantile men in relation to the highest interests of those working under them and with them is proverbial. Of late years, however, it has often been found possible to awaken even such bodies by strong and rightly-timed appeals to beneficent and ameliorative action. Though many coal-owners individually effect much good, yet collectively, we fear, they are as apathetic as any commercial body. Scattered over different districts, widely separated from each other, seldom coming into personal communication, and always deeply occupied in their callings and collieries, it is difficult to devise any method of appealing to and influencing them for good—for higher aims, for the attainment of the greatest efficiency and security in their collieries, and the steady advancement of their men in education, in mining ability, and in commendable conduct. For their own professional protection they can form a Mining Association which presents an united front; it is much to be desired that they would present equal unity and energy against ignorance, incompetence, and carelessness. Mining in the best coal-seams is yearly becoming more difficult, and yearly requires increased ability and fertility of practical resources in managers. Men of mining mark are frequently tempted away from the most systematically wrought pits to other districts by higher remuneration. Mr. Wood declared twelve years ago, that there was in

[1] An Educational Institute has recently been opened at Newcastle-upon-Tyne, and an appropriate and handsome building is about to be erected at a cost of 4,000*l.*, if that sum can be obtained. This building will include the Coal Trade, Mining Institute, and the Wood Memorial Hall. The family of Mr. Wood have liberally assisted this enterprise, but it seems strange indeed that so wealthy a body as the coal-owners of the North of England do not at once contribute 4,000*l.*

his vicinity in the Durham district, an actual scarcity of properly educated persons to take the management of mines. The class of competent managers has not been largely and proportionably increased since that time. The great body of pit-working people is now in danger of decreasing as well as the small body of fully educated superintendents. Were there good local mining colleges or high schools, this latter class would be gradually augmented. Such establishments are urgently required to raise the tone, form the character, and develope the talents of the young men who now live in and around the large coalfields, with little stimulus to, and few means for, self-improvement. In such districts as South Staffordshire and Worcestershire even the overseers are notoriously ignorant, so much so that the local inspector, Mr. Baker, states in evidence —' We occasionally find the overseers of the pits quite as ignorant as the working-men. They are good men for all practical purposes in the mine, but as for education some of them are very ignorant indeed.'

We have purposely refrained from observations on the two notorious recent explosions in Yorkshire and Staffordshire, because the daily newspapers have published nearly all the evidence taken at the inquests, and because, also, they form the special subjects of reports to the Home Secretary, which will be shortly made public. The tenor of these reports may be easily anticipated, and what we shall learn from them will in all probability only be a more authentic record of the evidence we already possess, and a statement of conclusions we may already foresee. The Select Committee, now reinstated, and we trust reinvigorated and stimulated by a lively consciousness of increased responsibility, arising from the terrible explosions which have occurred, as it were, in the midst of their deliberations, will doubtless devote special attention to this subject. It will be vexatious indeed, if from all these hopeful quarters, in addition to the accumulated labours, reports, and evidence of previous committees who have examined able and abundant witnesses relating to mining accidents, and of commissioners who have personally visited the pits, and have reported upon them in great detail, we do not ultimately obtain a satisfactory system for the benefit of the 315,000 coal-miners working in the inspected districts. With all the science, practical skill, and appliances at our command, it will be indeed distressing if these hard-working men should still continue to lose on the average one life for every 321 persons employed. A death on every working day in the year is surely not the ultimate attainable minimum of coal-mining mortality.

APPENDIX.

In this Appendix I proceed to offer some additional and confirmatory information upon the following subjects: —
I. The Output of Coal during the year 1872, and the number of persons then employed in our coal mines.
II. Colliers' Wages and Work. III. The Rise in the Price of Coal. IV. The Supply, Consumption, and Cost of Coal in the future.

I. *The Output of Coal in the Year* 1872, *and the Number of Persons then Employed in our Coal Mines.*

The following is a tabular view of the output in the United Kingdom during last year as compiled from the evidence of the Inspectors of Mines, and printed in *The Mining Journal* of May 3rd :—

	Tons, 1872.
South Durham	17,395,000
Northumberland, Durham, &c. . .	13,000,000
Yorkshire	14,536,000
Derbyshire, &c.	10,660,000
Lancashire and North Wales . . .	18,363,236
North Staffordshire	16,877,188
Gloucester and Somerset . . .	7,000,000
South Wales	10,131,725
Scotland	15,383,609
Ireland (estimate for 1872) . . .	200,000
	123,546,758

The next tabular view shows the number of persons employed in each of the districts specified, with the average quantity in tons raised by each person during the year 1872.

	Persons employed in 1872.	Average of tons.
South Durham . . .	45,300	384
Northumberland, &c. . .	39,000	333
Yorkshire	51,056	285
Derbyshire, &c. . . .	39,200	271
North Staffordshire . . .	27,555	228
South Staffordshire . . .	31,500	335
Lancashire, West . . .	28,657	326
Lancashire, North, &c. . .	34,000	261
Gloucester, &c. . . .	27,300	256
South Wales	38,427	263
Scotland, West . . .	20,639	307
Scotland, East	30,000	301

II. *Colliers' Wages and Work.*

From the evidence already given before the House of Commons' Committee on coal, it appears that the rise in colliers' wages or earnings has been very considerable in all districts, especially in the case of the hewers. With reference to Lancashire, Mr. Alfred Hewlett, who is Managing Director of the Wigan Coal Company, and both a colliery viewer and proprietor, and who further has about 10,000 men employed in various pits, stated that the average of one man during a fortnight in April of this year was 24s. 1d. per day; of another, the average was 26s. 10d.; of another it was 9s. 6d.; of another, 9s. 10d.; of another it was 10s. 4d.; of another, 9s. 11d.; of another, 17s. 9d.; of another, 12s. 10d.; of another, 16s. 11d., in all instances, after deducting what the men are allowed. Such were the actual daily earnings of hewers, who worked from eight to eight and a half days in the fortnight. It would have been in the power of these men to earn this rate of wages every working day in the fortnight, and their employers would have been very glad if they had so worked. The men could have worked the whole twelve days without injury to their health. On the whole, the wages have increased 64¾ per cent. from September, 1871 to 1873; and of this 50 per cent. is due to 1872.

The above seems to be the most remarkable instance of a great increase of earnings, but in other districts a considerable rise has also followed. The Inspector of Mines for Lanarkshire and adjacent parts in Scotland stated, that in his district wages have greatly increased. Labour is scarce now, and they cannot get Irish labourers as formerly. The preceding prices refer to hewers and getters of coal, but unskilled labourers who used to receive 3s. 6d. and 4s. a day, now obtain from 8s. to 10s. a day. The Inspector of a neighbouring district to the above-named, said that for the four years preceding 1872 the wages of miners were somewhere about 4s. 2d. a day, whereas last year the average was nearer 7s. 6d. a day—understood wage. In another Scotch district, where the colliers work by piece, the work is so set out that by a fair day's work the men can now make 7s. 6d. a day.

Mr. Booth, the General Manager of the Claremont and other collieries at Ashton-under-Lyne, and who has about 1,600 men in his district, stated that in his two principal mines the wages are, in the one case, 45 per cent. greater than in 1867, and 156 per cent. more than in March, 1869. In the other case, the amount is 47 per cent. greater than in 1867, and 161 per cent. more than in March, 1869. This calculation is founded on the price paid per ton, taking together the whole gang of colliers employed below ground in getting coal, and upon a division of the total wages by the number of days work at eight hours a day.

For the fortnight ending March 19th, 1873, the average earnings in the different mines were 8s. 6d., 8s. 11d., 10s. 3d., 10s. 9d., 9s. 9d., 8s. 8d., 9s. 1d., and 9s., the general average being 9s. 3d., reckoning men and boys together.

Mr. Booth thinks the men have profited pecuniarily by the rise in wages, that is, they have received more money, though they have not worked so long. Men who understood their work,

and worked well and continually, could earn large sums. In
one instance, two men and a boy have not for a considerable
period taken less than 22*l.* a fortnight. But generally the men
now work fewer turns in a fortnight, at the rate of nine as
against eleven formerly, so that they have lost two days a fortnight
upon the general average since the rise in wages.

The same witness pronounced his opinion that by far the
majority of men spend the money as fast as they get it, and that
the principal portion of them are even worse off than before.
Though they make more money, they are in social condition
much worse than before. The men combined, in June last, to
work no more than eight hours a day. The men earn as much
as before, but the consumers have to pay a higher price for their
coal. Many of them take every Monday and Tuesday for play,
and then work 20 per cent. more on the Wednesday and Thurs-
day forming the last two days in the fortnight. Neither are their
houses better furnished, nor are their families better clothed, for
the rise in wages.

Mr. Hewlett declared that the men have been more and more
disinclined to work full time. In the pits he supervises at Wigan,
which are now working full time as against half time in 1869,
there has been a deficiency in the output arising from this cause.
The working hours are one and a half to two less per day, and
the number of days worked have also been much reduced. The
average amount of work per man per fortnight is not more than
from eight to eight and a half days. In the beginning of the
year the men asked to have their wages paid weekly, instead of
fortnightly, and so anxious were the coal owners to increase the
output, that they acceded to their request if the men would
agree to work every day in the fortnight for six months, and
consent to forego the ' play Monday.' Upon the consent of the
men having been obtained, the owners were put to considerable
expense in keeping the pits in full working order, but very few
men indeed have worked full time.

The same gentleman expressed his opinion that the alterations
which have been made in the hours of labour of young persons
between twelve and sixteen, have caused a very serious check to
be placed upon the output of coal in conjunction with the disin-

-clination of men to work full hours, and that the effect of this check will be to decrease the output by 15 per cent.

Under the new Act of Parliament, the hewers cannot work beyond fifty-four hours in one week, supposing they have the assistance of young persons. If the hewer were disposed to work regularly and fully for fifty-four hours in the week, he would probably produce a great deal more coal than at present; but, in fact, he is paid on Saturday night, does not come on Monday, and very likely not on Tuesday, and perhaps only does part of a day's work on Wednesday, so that it is really only in the latter part of the week that he gets into full swing. In former times, he used to make up in the second week what he lost in the first.

As far as Mr. Hewlett could judge in the last three months, the effect of the restrictions and conditions just named, has been to increase the cost of coal by 1s. 4d. to 1s. 8d. per ton.

Mr. Booth also thought that the operation of the Mines Regulation Act would be to diminish the supply of colliers, since employers cannot now have the same number of boys as formerly, and they cannot be worked as they once were. Besides, it is very difficult to get them to go underground at all, unless they go at an early age, and mothers have a great objection to their sons going down if they can get them anything else to do. No other labourer would become so good a collier as if he had gone down when a boy. If boys cannot be got to work in the pits, this circumstance alone will restrict the output; nor is any great increase of output to be looked for, notwithstanding the great rise in wages.

The manager of the Earl of Dudley's collieries, in South Staffordshire, which employ from 8,000 to 10,000 men, observed that the pits in his district were able to put out more coal than they do if more men could be obtained, or if the men at present employed would work more time. He agreed with Mr. Booth with respect to the want of improvement in the men, and their homes and habits, by the rise in wages; and, as a magistrate, grieved to say the number of cases of drunkenness have very much increased. There will be a great scarcity of good colliers in future, owing to the circumstances above named; and, in fact, in South Staffordshire, the parents are taking all the boys out of the pits, and have been gradually doing this for the last seven or eight years.

For West Yorkshire, Mr. Tennant, of Leeds, affirmed that many hewers make £4 a week, and that some few take home about £5 a week, who work six days a week to earn this sum. He had not observed any great difference in the men in consequence of the. advance in their wages, and no general social improvement. The price of the labour alone, as between 1867 and 1873, has risen from 2s. 10d. per ton to 6s. 5d. (This statement gave rise to numerous questions.)

III. *Rise in the Price of Coal.*

In West Lancashire the rise in the price of coal has been about 75 per cent., and while coal has there been rising, a considerable exportation of it has been going on from Lancashire by sea. In the western district of Scotland, the inspector paid 10s. 6d. for 24 cwt. of coal for domestic consumption in the year 1860; but in 1872 he had to pay 28s. 6d. for the same quantity, though he only paid 12s. or 14s. in 1871. He attributed this rise to the great excess of demand over supply in pig iron, the price of which has gone up from 50s., at which it stood for a number of years, to 6l. and 8l. per ton.

In South Staffordshire the greatest demand for coal was admitted to be for ironworks.

Mr. Booth, of Ashton-under-Lyne, attributed the rise in price to the greatly increased demand for coal for manufacturing purposes of all descriptions, but chiefly for cotton and iron manufactories; not so much in his district for iron making as for the manufacturing of cotton. The machine shops largely demanded coal; and the increased number of mills and workshops, all of which used gas, rendered more coal necessary for gas making to meet the demand. The demand for coal gradually increased up to the year 1873, until, in January last, everybody got alarmed while coals were not to be had, and the enormous demands could not, by any working, be met. Mr. Booth does not think that the masters can combine to keep up the prices, though the action of the men can and does raise the price; which is largely influenced by the rate of wages.

The manager of the Earl of Dudley's collieries, in South Staffordshire, gave evidence much to the same effect. The shortening of the hours of labour has seriously interfered with extraction of coal. In 1872 the coal averaged from 11*s*. 3*d*. to 15*s*. per ton; in 1873 from 15*s*. to 19*s*., the latter being the present price. Coal began to rise in 1869; it got to 10*s*. 3*d*. in September 1871, and gradually to 19*s*. He believed that the Association of Coalowners, of which he was the Chairman, 'could have put the price up to a very great extent, during the last year or two.'

Several of the witnesses confirm the fact, that, from 1870 up to the end of 1872, the output of coal has decreased, while the demand has increased, so that old stocks were speedily drawn upon and by the end of 1872 were altogether exhausted. For instance, according to Mr. Hewlett, his company, at Wigan, had 300,000 tons of coal in stock towards the latter part of 1870; and there were other large stores in hand at other places. As long as there were such stocks to fall back upon, the diminution of the output was not felt, but when an increased demand in manufacturing and domestic consumption sprang up, and the old stocks were exhausted, naturally the price of coal increased to the rates we have been lately familiar with.

The increased rate of wages is connected with this, and it is evident that the high price of coal has led to an increased demand for high wages.

Another circumstance affecting the cost of coal, is the custom of some districts to contract very largely for the delivery of certain quantities at certain prices. Probably 60 per cent. of the coal produced in the district round Wigan, has been raised to supply under contract. Mr. Hewlett thinks that there are a very large number of contracts in force at present which cause an actual loss. Perhaps they were made in or before 1869, when wages and materials were far less in price.

The difference in the prices of coal raised since 1867 is thus

stated by Mr. Hewlett:—In the year 1868 it was 6½ per cent.
less than in 1867; in 1869 it was 9⅓ per cent. less; in 1870
it was still 9½ per cent. less; in 1871 it was eight per cent. less;
in 1872 it was 34 per cent. more than in 1867; and in the first
quarter of 1873 it was 82½ per cent. higher. These differences
are calculated upon coal up to the period of its being put in the
trucks on the sidings to be carried away.

Mr. Robert Tennant, of Leeds, a coal owner and a member of
the West Yorkshire Colliery Proprietors' Association, is familiar
with that district, and supposes there are about 300 collieries in it.

The tonnage raised last year amounted to 7,000,000, and
there are about 25,000 men and boys altogether employed in the
various collieries. He gave evidence that the price of coal in
West Yorkshire has been raised since 1870 from 75 to 150 per
cent. to meet the increased demand. For the years from 1867
to 1870 inclusive, the average selling price in his own colliery
was 5s. 3d. per ton, that is for every variety of coal brought to
the surface. During the year 1871 the average selling price for
all sorts of coal was 5s. 8d. per ton, and in 1872 it was 9s. 3d.
per ton; the great increase in 1872 having taken place in the
latter part of the year. The price of the very best coal is now
about 20s. a ton. Four years ago it may have been about
9s. 6d. For the entire district it would be no exaggeration to say
that the best coal has risen in cost by from 75 to 100 per cent.

Mr. Tennant was closely pressed respecting the power of the
Colliery Proprietors' Association to put up the price of coal, and
replied: ' I will not be certain that the price of coal would have increased as it did, unless there had been some sort of arrangement
amongst coal owners that it should rise. There was an attempt
made of which I entirely disapproved, and I ceased to be Chairman.'

This gentleman was urged to produce the pay-sheets of his
colliery, in order to show the sums received per man, as, said
Mr. Alderman Carter, ' the colliers deny entirely that such
wages as you state are paid in the district.' Mr. Tennant had
no personal objection to produce them and promised to do so.

The above-cited testimony will sufficiently explain the views
entertained by some of the best-informed witnesses examined
before the Committee.

IV. *The Supply, Consumption, and Cost of Coal in the Future.*

It is natural that after treating of the preceding topics we should turn an anxious eye towards the future; and the distinction drawn in the first Article between causes in continuous, and causes in temporary operation, with relation to consumption and cost, will materially help us to form a conjectural opinion upon a matter interesting to all.

It has been established incontrovertibly that amongst the causes in continuous operation to enhance cost, the principal one is the revival and prosperity of iron-making and metalliferous trades, as well as of cotton-spinning and other manufactures. If these trades and enterprises increase in prosperity, the cause on their side will be constantly and simultaneously operative. The more pig iron, railway iron, angle iron, and copper we make, the more of coal we shall require and consume. In fact, the demand and rise in iron and coal are inseparably associated. Should these enterprises and trades decline, the cost of coal will in like proportion diminish.

It is true that the best household coal is not of the same kind as that consumed in ordinary manufactories, but in some hitherto unexplained manner the price of the best household coal always rises with that of inferior coal. Whether this is an optional or a necessary consequence, it is impossible to decide. Probably any rise in one branch stimulates the desire to raise the price in all other branches. This may be owing to collusion or not, but the result to the public is the same. It is doubtful if the Metropolitan public have really been receiving real Newcastle, or Hetton coal, at all during the recent coal famine. A portion of the Northern household coal has been lately diverted to iron works.

Some other causes will not only be continuously, but also increasingly operative. One of these is the demand for gas, which will be concurrent with increase of population and the extension of large towns and cities. Nearly all the Gas Companies are prosperous, and, notwithstanding several projects of an experimental character, no substitute for gas distilled for the destruction of coal has yet been successful to any large extent. The best coal is essential to the manufacture of gas, and the well-known Cannel Coal of Wigan, in Lancashire, is the most appropriate. In Germany it has been found that the Brown Coal, or imperfect Lignite, is very useful in producing a certain kind of gas; but as we have very little coal of the same kind in Great Britain, this discovery does not concern us. It concerns Berlin, and there will probably be of some importance. We have still to depend upon the best gas coal for a largely increasing demand. Every house, mill, and workshop illuminated with gas, is to that extent an item in the increased demand for gas coal.

In the first Article in this book, I endeavoured to assign the proportionate values to the several distinct causes for demand of coal. I now add for illustration the probable distribution of every thousand tons of coal raised in the kingdom, as far as it can be approximately represented:—

As it was observed by Mr. I. Lowthian Bell in his recent Presidential Address at the Iron and Steel Institute, roughly, according to the figures in the Reports of the Royal Commissioner, every thousand tons raised is disposed of as follows:— In paper-making and preparing, 6; in smelting copper, lead, tin, and zinc, 8; in waterworks, 14; in breweries and distilleries, 18; in chemical manufactures, 19; in railway works, 20; in steam navigation, 30; in works of clay and glass and lime-kilns, 31; in textile fabrics—wool, cotton, silk, flax, and jute, 42; in gas works, 60; in mining operations, 67; in coal exported to foreign countries, 92; in general purposes, chiefly steam engines, 121; in domestic use, 172; in iron and steel works, inclusive of coal required for their steam power, 300; $=1,000$ in the total. Hence it appears that the iron trade of this country requires as much coal as any of the two largest sources of consumption.

This statement completely establishes the view I have taken of the demands for coal for iron and metalliferous manufactures. When these are exceptionally prosperous, coal will become proportionately dear. Manufacturing supremacy is bounded by the supply and cost of coal, and that particular industry suffers most into which the cost of fuel most largely enters.

It is apparent that, assuming the household consumption of coal to be as 172 against 300 in iron and steel works, any small economy effected in household use will not diminish considerably the total demand. To some extent it will tell, but not materially, in a gross consumption of 123,000,000 of tons per annum.

The price of coal has already fallen considerably, namely, to 30s. per ton for best household, taking the price in the London Coal Exchange as quoted at the time of writing these lines; that is, it has fallen from an extreme panic price to an ordinarily dear price. Whether it will fall still lower, and whether the household consumer will gain proportionately when coal is delivered to him, are subjects of conjecture.

As regards the colliers, if they can contrive to combine and keep up their wages, while they diminish the total output, by just so much will they enhance the price of coal. Although the evidence at present given before the Select Committee is against them, it is possible that they may adduce counter evidence, and in all fairness they will be duly heard. Nevertheless, the existing evidence is decidedly against them, and if they dispute it, then it must be verified by an appeal to pay-sheets and authentic figures. No doubt if the demand decreases, their earnings will correspondingly decrease. They themselves are aware of this fact, and no effort of theirs can countervail it in the long run.

The increasing scarcity of hewers is a very important element in the consideration of the future cost of coal. I have already described the peculiarity of their work, and the evidence given shows that no men can do this work so effectively as those brought up to it from boyhood. This fact is well known to all who have inspected coal mines, and no attempts to attract agricultural or other labourers to pit work will entirely surmount the difficulty. Apparently there will be a great demand for good hewers, and these men cannot be immediately multiplied to the occasion. All other departments of colliery labour can be recruited *ad libitum*, but this one of hewing, especially in thin seams, is the skilled and difficult department of work. I need not add to the preceding observations on this topic.

Quite opposite opinions are expressed as to the possibility of an increased output of coal. It will be seen that some of the witnesses do not expect it, or consider it practicable, while Mr. I. Lowthian Bell says, ' in my opinion an important addition can and will be made to the present output, but it is very possible that the time is approaching when any extension of manufacturing operations in this country will have to be regulated, not by the requirements of society for their produce but by the means our coal mines possess of providing the fuel required.'

The late extraordinary demand for coal, beyond the output, has naturally led to a very active exploration for coal in all our coal fields, greater indeed than has been known for many years, and to a corresponding eagerness and enterprise in forming companies and commencing methods for developing the mineral. We are informed by *The Mining Journal*, May 3rd, that during the week vast fields of coal, that cannot be calculated by acres but by miles, have been bored and will be at once opened out. The prices paid are such as would not have been dreamt of a year ago. In many instances, 420*l.* per acre has been given for coal for which, not very long since, 300*l.* would have been considered too much. With the improved appliances now available, most of the new mines, it is expected, will be finished in about two years, so that, looking at the number of pits now being sunk, and those about to be commenced, the increase, during the next two or three years, of coal of every description, will be immense.

Numerous projects of this nature have come under our notice since the first Article was written. In the Midland Coal Field, which commences near Nottingham and extends to Leeds, embracing an area of more than 300 square miles, new collieries are being opened through its entire length. A very important sinking is also now about to be made by the Barrow Steel and Iron Company, from the Barnsley to the Silkstone coal, the depth between the two seams being about 380 yards, and the area of coal that is expected to be thereby worked extends for many miles in nearly all directions. The journals devoted to coal and mining are weekly bringing before their readers new schemes for exploration or declared discoveries of coal.

Making all allowance for the possible failure of many of these schemes and projects, it is still highly probable that the output of coal will be largely and progressively increased. The working colliers, at a recent meeting, pronounced their determination to enter for themselves into coal-mining, and it was referred to a committee to enquire and report upon the best method of following out their determination.

So far as these schemes prosper, so far a future 'coal-famine' appears to be unlikely, because, in the recent scarcity, temporary combined with continued causes. Thus, then, it is probable that cost can be materially lessened, at least for domestic consumption.

In reference to the greatest sources of consumption, I need not repeat previous observations and figures, but will only add that, considering the make of pig iron in Great Britain in 1862 was about 4 millions of tons, and that in 1872 it was probably 6¾ millions of tons, the increased consumption of coal in ten years, for this make alone, has been an important continuous cause of the rapidly progressive demand for coal; and, as Mr. I. L. Bell states, the make of pig iron ought, at the past ratio of increase, to reach 11½ millions of tons in the year 1882. When, for this, and for the probable extension of ironworks, something like 65 millions of tons of coal may be required, I do not think I have taken too alarming a view of the future in my first article on the consumption and cost of coal.

We shall doubtless have numerous and possibly rapid fluctuations in price; but if the great and largely increasing consumers of coal prosper, the demand for coal must correspondingly increase, and such results as I have briefly indicated must almost inevitably follow. Naturally, very different opinions are entertained on this subject, but my principal object has been to enable my readers to form their own opinion upon the facts, statistics, and data, which have led me to arrive at mine.

LONDON: PRINTED BY
SPOTTISWOODE AND CO., NEW-STREET SQUARE
AND PARLIAMENT STREET

Traveller's Library Editions,

PRICE ONE SHILLING EACH,

LEGIBLY PRINTED AND SUITABLE FOR SCHOOL PRIZES.

OUR COAL-FIELDS and OUR COAL-PITS. Two Parts, 1s. each; or in 1 vol. price 2s. 6d. cloth.

WARREN HASTINGS. By Lord Macaulay.

LORD CLIVE. By Lord Macaulay.

WILLIAM PITT and the EARL of CHATHAM. By Lord Macaulay.

LORD MACAULAY'S ESSAYS on RANKE'S HISTORY of the POPES, and GLADSTONE on CHURCH and STATE.

LORD 'MACAULAY'S ESSAYS on ADDISON and WALPOLE.

LORD BACON. By Lord Macaulay.

LORD MACAULAY'S ESSAYS on LORD BYRON and the COMIC DRAMATISTS of the RESTORATION.

LORD MACAULAY'S ESSAY on FREDERICK the GREAT.

LORD MACAULAY'S ESSAY on HALLAM'S CONSTITUTIONAL HISTORY of ENGLAND.

LORD MACAULAY'S ESSAY on DR. SAMUEL JOHNSON.

LORD MACAULAY'S SPEECHES on PARLIAMENTARY REFORM.

SIR ROGER DE COVERLEY. From the *Spectator*.

SIR EDWARD SEAWARD'S NARRATIVE of his SHIP-WRECK. Two Parts, 1s. each; or in 1 vol. price 2s. 6d. cloth.

SWISS MEN and SWISS MOUNTAINS. By R. Ferguson.

An ATTIC PHILOSOPHER in PARIS. By E. Souvestre.

London: LONGMANS and CO.

ENCYCLOPÆDIAS AND DICTIONARIES.

M'CULLOCH'S DICTIONARY OF COMMERCE AND
COMMERCIAL NAVIGATION. New Edition, revised and corrected
throughout; with Supplements containing Notices or Abstracts of the New
Tariffs for the United States and Spain, and of our New Bankruptcy,
Naturalisation, and Neutrality Laws. Edited by HUGH G. REID. With
Eleven MAPS and THIRTY CHARTS. In One Volume, medium 8vo. price 63s.

OATES'S NEW DICTIONARY OF GENERAL BIO-
GRAPHY; Containing Concise Memoirs and Notices of the most Eminent
Persons of all Countries, from the Earliest Ages to the Present Time. With
a Classified and Chronological Index of the Principal Names. In One
Volume, medium 8vo. price 21s.

CATES AND WOODWARD'S ENCYCLOPÆDIA OF
CHRONOLOGY, HISTORICAL AND BIOGRAPHICAL; Comprising the
DATES of all the Great Events of History, including Treaties, Alliances,
Wars. Battles, &c.; Incidents in the lives of Eminent Men, and their Works,
Scientific and Geographical Discoveries, Mechanical Inventions, and Social,
Domestic and Economical Improvements. Medium 8vo. pp. 1496 (double
columns, brevier), price 42s. cloth.

KEITH JOHNSTON'S GENERAL DICTIONARY OF
GEOGRAPHY. Descriptive, Physical, Statistical, and Historical. Forming a Complete Gazetteer of the World. New and improved Edition, carefully revised throughout. In One Volume, medium 8vo. [*In the Press.*

URE'S DICTIONARY OF ARTS, MANUFACTURES,
AND MINES. Sixth Edition, rewritten and enlarged, by ROBERT HUNT,
F.R.S. Keeper of Mining Records; assisted by numerous Contributors
eminent in Science and familiar with Manufactures. With about 2,600
Engravings on Wood. In Three Volumes, medium 8vo. price £4. 14s. 6d.

A DICTIONARY OF CHEMISTRY AND THE ALLIED
BRANCHES OF OTHER SCIENCES, founded on that of the late Dr.
Ure. By HENRY WATTS. B.A. F.R.S. F.C.S. assisted by Eminent Scientific
and Practical Chemists. In FIVE VOLUMES, medium 8vo. price £7. 3s.

SUPPLEMENTARY VOLUME of Recent Chemical Discoveries. Price 31s. 6d.

COPLAND'S DICTIONARY OF PRACTICAL MEDICINE,
Abridged from the larger work by the Author, assisted by J. C. COPLAND,
M.R.C.S. and throughout brought down to the present state of Medical
Science. In One Volume, medium 8vo. price 36s.

GWILT'S ENCYCLOPÆDIA OF ARCHITECTURE.
Illustrated with more than 1,100 Engravings on Wood, revised, with Alter-
ations and considerable Additions, by WYATT PAPWORTH. Fellow of the
Royal Institute of British Architects. Additionally Illustrated with nearly
400 Engravings on Wood by O. Jewitt; and more than 100 other Woodcuts.
In One Volume, medium 8vo. price 52s. 6d.

BRANDE'S DICTIONARY OF SCIENCE, LITERATURE,
AND ART; Comprising the Definitions and Derivations of the Scientific
Terms in general use, together with the History and Description of the
Scientific Principles of nearly every Branch of Human Knowledge. Fourth
Edition, reconstructed by the late Author and the Rev. G. W. COX, M.A.
assisted by Contributors of eminent Scientific and Literary Acquirements.
In Three Volumes, medium 8vo. price 63s.

London : LONGMANS and CO. Paternoster Row.

[MARCH 1873.]

GENERAL LIST OF WORKS

PUBLISHED BY

MESSRS. LONGMANS, GREEN, AND CO.

PATERNOSTER ROW, LONDON.

History, Politics, Historical Memoirs, &c.

The HISTORY of ENGLAND from the Fall of Wolsey to the Defeat of the Spanish Armada. By JAMES ANTHONY FROUDE, M.A. late Fellow of Exeter College, Oxford.
> LIBRARY EDITION, 12 VOLS. 8vo. price £3. 18*s*.
> CABINET EDITION, in 12 vols. crown 8vo. price 72*s*.

The ENGLISH in IRELAND in the EIGHTEENTH CENTURY. By JAMES ANTHONY FROUDE, M.A. late Fellow of Exeter College, Oxford. VOL. I 8vo. price 16*s*.

ESTIMATES of the ENGLISH KINGS from WILLIAM the CON-QUEROR to GEORGE III. By J. LANGTON SANFORD. Crown 8vo. price 12*s*. 6*d*.

The HISTORY of ENGLAND from the Accession of James II. By Lord MACAULAY.
> STUDENT'S EDITION, 2 vols. crown 8vo. 12*s*.
> PEOPLE'S EDITION, 4 vols. crown 8vo. 16*s*.
> CABINET EDITION, 8 vols. post 8vo. 48*s*.
> LIBRARY EDITION, 5 vols. 8vo. £4.

LORD MACAULAY'S WORKS. Complete and Uniform Library Edition. Edited by his Sister, Lady TREVELYAN. 8 vols. 8vo. with Portrait, price £5. 5*s*. cloth, or £8. 8*s*. bound in tree-calf by Rivière.

VARIETIES of VICE-REGAL LIFE. By Sir WILLIAM DENISON, K.C.B. late Governor-General of the Australian Colonies, and Governor of Madras. With Two Maps. 2 vols. 8vo. 28*s*.

On PARLIAMENTARY GOVERNMENT in ENGLAND; its Origin Development, and Practical Operation. By ALPHEUS TODD, Librarian of the Legislative Assembly of Canada. 2 vols. 8vo. price £1. 17*s*.

The CONSTITUTIONAL HISTORY of ENGLAND, since the Accession of George III. 1760—1860. By Sir THOMAS ERSKINE MAY, C.B. Second Edition. Cabinet Edition, thoroughly revised. 3 vols. crown 8vo. price 18*s*.

A

The **HISTORY** of **ENGLAND**, from the Earliest Times to the Year 1865. By C. D. YONGE, B.A. Regius Professor of Modern History in Queen's College, Belfast. New Edition. Crown 8vo. price 7s. 6d.

The **OXFORD REFORMERS**—John Colet, Erasmus, and Thomas More; being a History of their Fellow-work. By FREDERIC SEEBOHM. Second Edition, enlarged. 8vo. 14s.

LECTURES on the **HISTORY** of **ENGLAND**, from the earliest Times to the Death of King Edward II. By WILLIAM LONGMAN. With Maps and Illustrations. 8vo. 15s.

The **HISTORY** of the **LIFE** and **TIMES** of **EDWARD** the **THIRD**. By WILLIAM LONGMAN. With 9 Maps, 8 Plates, and 16 Woodcuts. 2 vols. 8vo. 28s.

MEMOIR and **CORRESPONDENCE** relating to **POLITICAL OCCUR-RENCES** in June and July 1834. By EDWARD JOHN LITTLETON, First Lord Hatherton. Edited, from the Original Manuscript, by HENRY REEVE. 8vo. price 7s. 6d.

WATERLOO LECTURES; a Study of the Campaign of 1815. By Colonel CHARLES C. CHESNEY, R.E. late Professor of Military Art and History in the Staff College. New Edition. 8vo. with Map, 10s. 6d.

ROYAL and **REPUBLICAN FRANCE**. A Series of Essays reprinted from the *Edinburgh, Quarterly*, and *British and Foreign Reviews*. By HENRY REEVE, C.B. D.C.L. 2 vols. 8vo. price 21s.

CHAPTERS from **FRENCH HISTORY**; St. Louis, Joan of Arc, Henri IV. with Sketches of the Intermediate Periods. By J. H. GURNEY, M.A. New Edition. Fcp. 8vo. 6s. 6d.

The **LIFE** and **TIMES** of **SIXTUS** the **FIFTH**. By Baron HÜBNER. Translated from the Original French, with the Author's sanction, by HUBERT E. H. JERNINGHAM. 2 vols. 8vo. price 24s.

IGNATIUS LOYOLA and the **EARLY JESUITS**. By STEWART ROSE. New Edition, revised. 8vo. with Portrait, price 16s.

The **SIXTH ORIENTAL MONARCHY**: or, the Geography, History, and Antiquities of Parthia. Collected and Illustrated from Ancient and Modern sources. By GEORGE RAWLINSON, M.A. Camden Professor of Ancient History in the University of Oxford. With Maps and Illustrations. 8vo. price 16s.

The **HISTORY** of **GREECE**. By C. THIRLWALL, D.D. Lord Bishop of St. David's. 8 vols. fcp. 8vo. price 28s.

GREEK HISTORY from Themistocles to Alexander, in a Series of Lives from Plutarch. Revised and arranged by A. H. CLOUGH. New Edition. Fcp. with 44 Woodcuts, 6s.

The **TALE** of the **GREAT PERSIAN WAR**, from the Histories of Herodotus. By GEORGE W. COX, M.A. New Edition. Fcp. 3s. 6d.

The **HISTORY** of **ROME**. By WILLIAM IHNE. English Edition, translated and revised by the Author. VOLS. I. and II. 8vo. price 30s.

HISTORY of the **ROMANS** under the **EMPIRE**. By the Very Rev. C. MERIVALE, D.C.L. Dean of Ely. 8 vols. post 8vo. 48s.

The FALL of the ROMAN REPUBLIC; a Short History of the Last
Century of the Commonwealth. By the same Author. 12mo. 7s. 6d.

THREE CENTURIES of MODERN HISTORY. By CHARLES DUKE
YONGE, B.A. Regius Professor of Modern History and English Literature in
Queen's College, Belfast. Crown 8vo. price 7s. 6d.

A STUDENT'S MANUAL of the HISTORY of INDIA, from the
Earliest Period to the Present. By Colonel MEADOWS TAYLOR, M.R.A.S.
M.R.I.A. Crown 8vo. with Maps, 7s. 6d.

The HISTORY of INDIA, from the Earliest Period to the close of Lord
Dalhousie's Administration. By JOHN CLARK MARSHMAN. 3 vols. crown
8vo. 22s. 6d.

INDIAN POLITY: a View of the System of Administration in India.
By Lieutenant-Colonel GEORGE CHESNEY, Fellow of the University of
Calcutta. New Edition, revised; with Map. 8vo. price 21s.

A COLONIST on the COLONIAL QUESTION. By JEHU MATHEWS,
of Toronto, Canada. Post 8vo. price 6s.

The IMPERIAL and COLONIAL CONSTITUTIONS of the BRI-
TANNIC EMPIRE, including INDIAN INSTITUTIONS. By Sir EDWARD
CREASY, M.A. With 6 Maps. 8vo. price 15s.

REALITIES of IRISH LIFE. By W. STEUART TRENCH, Land Agent
in Ireland to the Marquess of Lansdowne, the Marquess of Bath, and Lord
Digby. Fifth Edition. Crown 8vo. price 6s.

The STUDENT'S MANUAL of the HISTORY of IRELAND. By
MARY F. CUSACK, Author of 'The Illustrated History of Ireland, from the
Earliest Period to the Year of Catholic Emancipation.' Crown 8vo. price 6s.

CRITICAL and HISTORICAL ESSAYS contributed to the *Edinburgh
Review*. By the Right Hon. LORD MACAULAY.

CABINET EDITION, 4 vols. post 8vo. 24s. | LIBRARY EDITION, 3 vols. 8vo. 36s.
PEOPLE'S EDITION, 2 vols. crown 8vo. 8s. | STUDENT'S EDITION, 1 vol. cr. 8vo. 6s.

SAINT-SIMON and SAINT-SIMONISM; a chapter in the History of
Socialism in France. By ARTHUR J. BOOTH, M.A. Crown 8vo. price 7s. 6d.

HISTORY of EUROPEAN MORALS, from Augustus to Charlemagne.
By W. E. H. LECKY, M.A. Second Edition. 2 vols. 8vo. price 28s.

HISTORY of the RISE and INFLUENCE of the SPIRIT of
RATIONALISM in EUROPE. By W. E. H. LECKY, M.A. Cabinet Edition,
being the Fourth. 2 vols. crown 8vo. price 16s.

GOD in HISTORY; or, the Progress of Man's Faith in the Moral
Order of the World. By Baron BUNSEN. Translated by SUSANNA WINK-
WORTH; with a Preface by Dean STANLEY. 3 vols. 8vo. price 42s.

The HISTORY of PHILOSOPHY, from Thales to Comte. By
GEORGE HENRY LEWES. Fourth Edition. 2 vols. 8vo. 32s.

An HISTORICAL VIEW of LITERATURE and ART in GREAT
BRITAIN from the Accession of the House of Hanover to the Reign of
Queen Victoria. By J. MURRAY GRAHAM, M.A. 8vo. price 12s.

The MYTHOLOGY of the ARYAN NATIONS. By GEORGE W.
COX, M.A. late Scholar of Trinity College, Oxford, Joint-Editor, with the
late Professor Brande, of the Fourth Edition of 'The Dictionary of Science,
Literature, and Art,' Author of 'Tales of Ancient Greece' &c. 2 vols. 8vo. 28s.

HISTORY of CIVILISATION in England and France, Spain and Scotland. By HENRY THOMAS BUCKLE. New Edition of the entire Work, with a complete INDEX. 3 vols. crown 8vo. 24s.

HISTORY of the CHRISTIAN CHURCH, from the Death of St. John to the Middle of the 2nd Century : comprising a full Account of the Primitive Organisation of Church Government and the Growth of Episcopacy. By the Rev. T. W. MOSSMAN, B.A. 8vo.   [Just ready.

HISTORY of the CHRISTIAN CHURCH, from the Ascension of Christ to the Conversion of Constantine. By E. BURTON, D.D. late Prof. of Divinity in the Univ. of Oxford. New Edition. Fcp. 3s. 6d.

SKETCH of the HISTORY of the CHURCH of ENGLAND to the Revolution of 1688. By the Right Rev. T. V. SHORT, D.D. Lord Bishop of St. Asaph. Eighth Edition. Crown 8vo. 7s. 6d.

HISTORY of the EARLY CHURCH, from the First Preaching of the Gospel to the Council of Nicæa. A.D. 325. By ELIZABETH M. SEWELL, Author of 'Amy Herbert.' New Edition, with Questions. Fcp. 4s. 6d.

The ENGLISH REFORMATION. By F. C. MASSINGBERD, M.A. Chancellor of Lincoln and Rector of South Ormsby. Fourth Edition, revised Fcp. 8vo. 7s. 6d.

MAUNDER'S HISTORICAL TREASURY; comprising a General Introductory Outline of Universal History, and a series of Separate Histories Latest Edition, revised and brought down to the Present Time by the Rev. GEORGE WILLIAM COX, M.A.   Fcp. 6s. cloth, or 10s. calf.

ENCYCLOPÆDIA of CHRONOLOGY, HISTORICAL and BIO- GRAPHICAL; comprising the Dates of all the Great Events of History, including Treaties, Alliances, Wars, Battles, &c.; Incidents in the Lives of Eminent Men and their Works, Scientific and Geographical Discoveries, Mechanical Inventions, and Social, Domestic, and Economical Improvements. By B. B. WOODWARD, B.A. and W. L. R. CATES. 8vo. price 42s.

Biographical Works.

BIOGRAPHICAL and CRITICAL ESSAYS. Reprinted from Reviews. with Additions and Corrections; a New Series. By A. HAYWARD, Q.C. 2 vols. 8vo. price 28s.

MEMOIR of GEORGE EDWARD LYNCH COTTON, D.D. Bishop of Calcutta and Metropolitan. With Selections from his Journals and Correspondence. Edited by Mrs. COTTON. Second Edition, with Portrait. Crown 8vo. price 7s. 6d.

MEMOIR of the LIFE of Admiral Sir EDWARD CODRINGTON. With Selections from his Public and Private Correspondence. Edited by his Daughter, Lady BOURCHIER. With Two Portraits, Maps, and Plans. 2 vols. 8vo. price 36s.

LIFE of Alexander VON HUMBOLDT. Compiled in Commemoration of the Centenary of his Birth. By Herr JULIUS LÖWENBERG, Dr. ROBERT AVÉ-LALLEMANT, and Dr. ALFRED DOVE. Edited by Professor KARL BRUHNS, Director of the Observatory at Leipzig. Translated by JANE and CAROLINE LASSELL. 2 vols. 8vo. with Three Portraits, price 36s.

MEMOIRS of BARON STOCKMAR. By his Son, Baron E. VON STOCKMAR. Translated from the German by G. A. M. Edited by F. MAX MÜLLER, M.A. 2 vols. crown 8vo. price 21s.

MUSICAL CRITICISM and **BIOGRAPHY**, from the Published and Unpublished Writings of THOMAS DAMANT EATON, late President of the Norwich Choral Society. Selected and edited by his SONS. Crown 8vo. 7s. 6d.

AUTOBIOGRAPHY of **JOHN MILTON**; or, Milton's Life in his own Words. By the Rev. JAMES J. G. GRAHAM, M.A. Crown 8vo. price 5s.

LORD GEORGE BENTINCK; a Political Biography. By the Right Hon. BENJAMIN DISRAELI, M.P. Eighth Edition, revised, with a New Preface. Crown 8vo. price 6s.

The **LIFE** of **ISAMBARD KINGDOM BRUNEL**, Civil Engineer. By ISAMBARD BRUNEL, B.C.L. of Lincoln's Inn; Chancellor of the Diocese of Ely. With Portrait, Plates, and Woodcuts. 8vo. 21s.

The **LIFE** and **LETTERS** of **FARADAY**. By Dr. BENCE JONES, Secretary of the Royal Institution. Second Edition, thoroughly revised. 2 vols. 8vo. with Portrait, and Eight Engravings on Wood, price 28s.

FARADAY as a **DISCOVERER**. By JOHN TYNDALL, LL.D. F.R.S. Professor of Natural Philosophy in the Royal Institution. New and Cheaper Edition, with Two Portraits. Fcp. 8vo. 3s. 6d.

RECOLLECTIONS of **PAST LIFE**. By Sir HENRY HOLLAND, Bart. M.D. F.R.S. &c. Physician-in-Ordinary to the Queen. Third Edition. Post 8vo. price 10s. 6d.

The **LIFE** and **LETTERS** of the **Rev. SYDNEY SMITH**. Edited by his Daughter, Lady HOLLAND, and Mrs. AUSTIN. New Edition, complete in One Volume. Crown 8vo. price 6s.

The **LIFE** and **TRAVELS** of **GEORGE WHITEFIELD**, M.A. By JAMES PATERSON GLEDSTONE. 8vo. price 14s.

LEADERS of **PUBLIC OPINION** in **IRELAND**; Swift, Flood, Grattan, and O'Connell. By W. E. H.'LECKY, M.A. New Edition, revised and enlarged. Crown 8vo. price 7s. 6d.

DICTIONARY of **GENERAL BIOGRAPHY**; containing Concise Memoirs and Notices of the most Eminent Persons of all Countries, from the Earliest Ages to the Present Time. Edited by W. L. R. CATES. 8vo. 21s.

LIVES of the **QUEENS** of **ENGLAND**. By AGNES STRICKLAND. Library Edition, newly revised; with Portraits of every Queen, Autographs, and Vignettes. 8 vols. post 8vo. 7s. 6d. each.

LIFE of the **DUKE** of **WELLINGTON**. By the Rev. G. R. GLEIG, M.A. Popular Edition, carefully revised; with copious Additions. Crown 8vo. with Portrait, 5s.

APOLOGIA PRO VITA SUA; being a History of his Religious Opinions, by JOHN HENRY NEWMAN, D.D., of the Oratory of St. Philip Neri. New Edition. Post 8vo. price 6s.

FELIX MENDELSSOHN'S LETTERS from *Italy and Switzerland*, and *Letters from 1833 to 1847*, translated by Lady WALLACE. New Edition, with Portrait. 2 vols. crown 8vo. 5s. each.

MEMOIRS of **SIR HENRY HAVELOCK**, K.C.B. By JOHN CLARK MARSHMAN. Cabinet Edition, with Portrait. Crown 8vo. price 3s. 6d.

VICISSITUDES of FAMILIES. By Sir J. BERNARD BURKE, C.B. Ulster King of Arms. New Edition, remodelled and enlarged. 2 vols. crown 8vo. 21s.

The **RISE of GREAT FAMILIES**, other Essays and Stories. By Sir J. BERNARD BURKE, C.B. Ulster King-of-Arms. Crown 8vo. price 12s. 6d.

ESSAYS in ECCLESIASTICAL BIOGRAPHY. By the Right Hon. Sir J. STEPHEN, LL.D. Cabinet Edition, being the Fifth. Crown 8vo. 7s. 6d.

MAUNDER'S BIOGRAPHICAL TREASURY. Thirteenth Edition, reconstructed, thoroughly revised, and in great part rewritten ; with about 1,000 additional Memoirs and Notices, by W. L. R. CATES. Fcp. 8vo. price 6s.

LETTERS and LIFE of FRANCIS BACON, including all his Occasional Works. Collected and edited, with a Commentary, by J. SPEDDING, Trin. Coll. Cantab. 6 vols. 8vo. price £3. 12s. To be completed in One more Volume.

Criticism, Philosophy, Polity, &c.

A SYSTEMATIC VIEW of the SCIENCE of JURISPRUDENCE. By SHELDON AMOS, M.A. Professor of Jurisprudence to the Inns of Court, London. 8vo. price 18s.

The **INSTITUTES of JUSTINIAN;** with English Introduction, Translation, and Notes. By T. C. SANDARS, M.A. Barrister, late Fellow of Oriel Coll. Oxon. New Edition. 8vo. 15s.

SOCRATES and the SOCRATIC SCHOOLS. Translated from the German of Dr. E. ZELLER, with the Author's approval, by the Rev. OSWALD J. REICHEL, B.C.L. and M.A. Crown 8vo. 8s. 6d.

The **STOICS, EPICUREANS, and SCEPTICS.** Translated from the German of Dr. E. ZELLER, with the Author's approval, by OSWALD J. REICHEL, B.C.L. and M.A. Crown 8vo. price 14s.

The **ETHICS of ARISTOTLE,** illustrated with Essays and Notes. By Sir A. GRANT, Bart. M.A. LL.D. Third Edition, revised and partly re-written.   [In the press.

The **NICOMACHEAN ETHICS of ARISTOTLE** newly translated into English. By R. WILLIAMS, B.A. Fellow and late Lecturer of Merton College, and sometime Student of Christ Church, Oxford. 8vo. 12s.

ELEMENTS of LOGIC. By R. WHATELY, D.D. late Archbishop of Dublin. New Edition. 8vo. 10s. 6d. crown 8vo. 4s. 6d.

Elements of Rhetoric. By the same Author. New Edition. 8vo. 10s. 6d. crown 8vo. 4s. 6d.

English Synonymes. By E. JANE WHATELY. Edited by Archbishop WHATELY. 5th Edition. Fcp. 3s.

BACON'S ESSAYS with ANNOTATIONS. By R. WHATELY, D.D. late Archbishop of Dublin. New Edition. 8vo. 10s. 6d.

LORD BACON'S WORKS, collected and edited by J. SPEDDING, M.A. R. L. ELLIS, M.A. and D. D. HEATH. New and Cheaper Edition. 7 vols. 8vo. price £3. 13s. 6d.

The **SUBJECTION of WOMEN.** By JOHN STUART MILL. New Edition. Post 8vo. 5s.

On REPRESENTATIVE GOVERNMENT. By JOHN STUART MILL. Third Edition. 8vo. 9s. Crown 8vo. 2s.

On LIBERTY. By JOHN STUART MILL. Fourth Edition. Post 8vo. 7s. 6d. Crown 8vo. 1s. 4d.

PRINCIPLES of POLITICAL ECONOMY. By the same Author. Seventh Edition. 2 vols. 8vo. 30s. Or in 1 vol. crown 8vo. 5s.

A SYSTEM of LOGIC, RATIOCINATIVE and INDUCTIVE. By the same Author. Eighth Edition. Two vols. 8vo. 25s.

UTILITARIANISM. By JOHN STUART MILL. Fourth Edition. 8vo. 5s.

DISSERTATIONS and DISCUSSIONS, POLITICAL, PHILOSOPHI-CAL, and HISTORICAL. By JOHN STUART MILL. Second Edition, revised. 3 vols. 8vo. 36s.

EXAMINATION of Sir W. HAMILTON'S PHILOSOPHY, and of the Principal Philosophical Questions discussed in his Writings. By JOHN STUART MILL. Fourth Edition. 8vo. 16s.

An OUTLINE of the NECESSARY LAWS of THOUGHT: a Treatise on Pure and Applied Logic. By the Most Rev. W. THOMSON, Lord Archbishop of York, D.D. F.R.S. Ninth Thousand. Crown 8vo. 5s. 6d.

PRINCIPLES of ECONOMICAL PHILOSOPHY. By HENRY DUNNING MACLEOD, M.A. Barrister-at-Law. Second Edition. In Two Volumes. VOL. I, 8vo. price 15s.

A Dictionary of Political Economy; Biographical, Bibliographical, Historical, and Practical. By the same Author. VOL. I. royal 8vo. 30s.

The ELECTION of REPRESENTATIVES, Parliamentary and Municipal; a Treatise. By THOMAS HARE, Barrister-at-Law. Fourth Edition, with Additions. Crown 8vo. 7s.

SPEECHES of the RIGHT HON. LORD MACAULAY, corrected by Himself. People's Edition, crown 8vo. 3s. 6d.

Lord Macaulay's Speeches on Parliamentary Reform in 1831 and 1832. 16mo. 1s.

A DICTIONARY of the ENGLISH LANGUAGE. By R. G. LATHAM, M.A. M.D. F.R.S. Founded on the Dictionary of Dr. SAMUEL JOHNSON, as edited by the Rev. H. J. TODD, with numerous Emendations and Additions. In Four Volumes, 4to. price £7.

THESAURUS of ENGLISH WORDS and PHRASES, classified and arranged so as to facilitate the Expression of Ideas, and assist in Literary Composition. By P. M. ROGET, M.D. New Edition. Crown 8vo. 10s. 6d.

LECTURES on the SCIENCE of LANGUAGE. By F. MAX MÜLLER, M.A. &c. Foreign Member of the French Institute. Sixth Edition. 2 vols. crown 8vo. price 16s.

MANUAL of ENGLISH LITERATURE, Historical and Critical. By THOMAS ARNOLD, M.A. New Edition. Crown 8vo. price 7s. 6d.

THREE CENTURIES of ENGLISH LITERATURE. By CHARLES DUKE YONGE, Regius Professor of Modern History and English Literature in Queen's College, Belfast. Crown 8vo. price 7s. 6d.

SOUTHEY'S DOCTOR, complete in One Volume. Edited by the Rev. J. W. WARTER, B.D. Square crown 8vo. 12s. 6d.

HISTORICAL and CRITICAL COMMENTARY on the OLD TESTA-
MENT; with a New Translation. By M. M. KALISCH, Ph.D. VOL. I.
Genesis, 8vo. 18*s.* or adapted for the General Reader, 12*s.* VOL. IL *Exodus*,
15*s.* or adapted for the General Reader, 12*s.* VOL. III. *Leviticus*, PART I.
15*s.* or adapted for the General Reader, 8*s.* VOL. IV. *Leviticus*, PART II.
15*s.* or adapted for the General Reader, 8*s.*

A DICTIONARY of ROMAN and GREEK ANTIQUITIES, with
about Two Thousand Engravings on Wood from Ancient Originals, illustrative of the Industrial Arts and Social Life of the Greeks and Romans.
By ANTHONY RICH, B.A. Third Edition, revised and improved. Crown 8vo.
price 7*s.* 6*d.*

A LATIN-ENGLISH DICTIONARY. By JOHN T. WHITE, D.D.
Oxon. and J. E. RIDDLE, M.A. Oxon. Third Edition, revised. 2 vols. 4to.
pp. 2,128, price 42*s.* cloth.

White's College Latin-English Dictionary (Intermediate Size),
abridged for the use of University Students from the Parent Work (as
above). Medium 8vo. pp. 1,048, price 18*s.* cloth.

White's Junior Student's Complete Latin-English and English-Latin
Dictionary. New Edition. Square 12mo. pp. 1,058, price 12*s.*

Separately { The ENGLISH-LATIN DICTIONARY, price 5*s.* 6*d.*
{ The LATIN-ENGLISH DICTIONARY, price 7*s.* 6*d.*

An ENGLISH-GREEK LEXICON, containing all the Greek Words
used by Writers of good authority. By C. D. YONGE, B.A. New Edition. 4to. 21*s.*

Mr. YONGE'S NEW LEXICON, English and Greek, abridged from
his larger work (as above). Revised Edition. Square 12mo. 8*s.* 6*d.*

A GREEK-ENGLISH LEXICON. Compiled by H. G. LIDDELL, D.D.
Dean of Christ Church, and R. SCOTT, D.D. Dean of Rochester. Sixth
Edition. Crown 4to. price 36*s.*

A Lexicon, Greek and English, abridged from LIDDELL and SCOTT's
Greek-English Lexicon. Fourteenth Edition. Square 12mo. 7*s.* 6*d.*

A SANSKRIT-ENGLISH DICTIONARY, the Sanskrit words printed
both in the original Devanagari and in Roman Letters. Compiled by
T. BENFEY, Prof. in the Univ. of Göttingen. 8vo. 52*s.* 6*d.*

A PRACTICAL DICTIONARY of the FRENCH and ENGLISH LAN-
GUAGES. By L. CONTANSEAU. Fourteenth Edition. Post 8vo. 10*s.* 6*d.*

Contanseau's Pocket Dictionary, French and English, abridged from
the above by the Author. New Edition, revised. Square 18mo. 3*s.* 6*d.*

NEW PRACTICAL DICTIONARY of the GERMAN LANGUAGE;
German–English and English-German. By the Rev. W. L. BLACKLEY, M.A.
and Dr. CARL MARTIN FRIEDLÄNDER. Post 8vo. 7*s.* 6*d.*

The MASTERY of LANGUAGES; or, the Art of Speaking Foreign
Tongues Idiomatically. By THOMAS PRENDERGAST, late of the Civil
Service at Madras. Third Edition. 8vo. 6*s.*

Miscellaneous Works and *Popular Metaphysics.*

MISCELLANEOUS and POSTHUMOUS WORKS of the Late HENRY
THOMAS BUCKLE. Edited, with a Biographical Notice, by HELEN
TAYLOR. 3 vols. 8vo. price 52*s.* 6*d.*

MISCELLANEOUS WRITINGS of JOHN CONINGTON, M.A. late
Corpus Professor of Latin in the University of Oxford. Edited by J. A.
SYMONDS, M.A. With a Memoir by H. J. S. SMITH, M.A. LL.D. F.R.S.
2 vols. 8vo. price 28s.

SEASIDE MUSINGS ON SUNDAYS AND WEEK-DAYS. By
A. K. H. B. Crown 8vo. price 3s. 6d.

Recreations of a Country Parson. By A. K. H. B. FIRST and
SECOND SERIES. crown 8vo. 3s. 6d. each.

The Common-place Philosopher in Town and Country. By A. K. H. B.
Crown 8vo. price 3s. 6d.

Leisure Hours in Town; Essays Consolatory, Æsthetical, Moral,
Social, and Domestic. By A. K. H. B. Crown 8vo. 3s. 6d.

The Autumn Holidays of a Country Parson; Essays contributed to
Fraser's Magazine, &c. By A. K. H. B. Crown 8vo. 3s. 6d.

The Graver Thoughts of a Country Parson. By A. K. H. B. FIRST
and SECOND SERIES, crown 8vo. 3s. 6d. each.

Critical Essays of a Country Parson, selected from Essays contributed to *Fraser's Magazine.* By A. K. H. B. Crown 8vo. 3s. 6d.

Sunday Afternoons at the Parish Church of a Scottish University
City. By A. K. H. B. Crown 8vo. 3s. 6d.

Lessons of Middle Age; with some Account of various Cities and
Men. By A. K. H. B. Crown 8vo. 3s. 6d.

Counsel and Comfort spoken from a City Pulpit. By A. K. H. B
Crown 8vo. price 3s. 6d.

Changed Aspects of Unchanged Truths; Memorials of St. Andrews
Sundays. By A. K. H.B. Crown 8vo. 3s. 6d.

Present-day Thoughts; Memorials of St. Andrews Sundays. By
A. K. H. B. Crown 8vo. 3s. 6d.

SHORT STUDIES on GREAT SUBJECTS. By JAMES ANTHONY
FROUDE, M.A. late Fellow of Exeter Coll. Oxford. 2 vols. cr. 8vo. price 12s.

LORD MACAULAY'S MISCELLANEOUS WRITINGS:—
LIBRARY EDITION. 2 vols. 8vo. Portrait, 21s.
PEOPLE'S EDITION. 1 vol. crown 8vo. 4s. 6d.

LORD MACAULAY'S MISCELLANEOUS WRITINGS and SPEECHES.
STUDENT'S EDITION, in crown 8vo. price 6s.

The Rev. SYDNEY SMITH'S MISCELLANEOUS WORKS; including his Contributions to the *Edinburgh Review.* Crown 8vo. 6s.

The Wit and Wisdom of the Rev. Sydney Smith; a Selection of
the most memorable Passages in his Writings and Conversation. 16mo. 3s. 6d.

The ECLIPSE of FAITH; or, a Visit to a Religious Sceptic. By
HENRY ROGERS. Twelfth Edition. Fcp. 8vo. price 5s.

Defence of the Eclipse of Faith, by its Author; a rejoinder to Dr.
Newman's *Reply.* Third Edition. Fcp. 8vo. price 3s. 6d.

CHIPS from a GERMAN WORKSHOP; being Essays on the Science
of Religion, and on Mythology, Traditions, and Customs. By F. MAX
MÜLLER, M.A. &c. Foreign Member of the French Institute. 3 vols. 8vo. £2.

ANALYSIS of the PHENOMENA of the HUMAN MIND. By JAMES MILL. A New Edition, with Notes, Illustrative and Critical, by ALEXANDER BAIN, ANDREW FINDLATER, and GEORGE GROTE. Edited, with additional Notes, by JOHN STUART MILL. 2 vols. 8vo. price 28s.

An INTRODUCTION to MENTAL PHILOSOPHY, on the Inductive Method. By J. D. MORELL, M.A. LL.D. 8vo. 12s.

ELEMENTS of PSYCHOLOGY, containing the Analysis of the Intellectual Powers. By the same Author. Post 8vo. 7s. 6d.

The SECRET of HEGEL: being the Hegelian System in Origin, Principle, Form, and Matter. By J. H. STIRLING, LL.D. 2 vols. 8vo. 28s.

SIR WILLIAM HAMILTON; being the Philosophy of Perception: an Analysis. By J. H. STIRLING, LL.D. 8vo. 5s.

LECTURES on the PHILOSOPHY of LAW. Together with Whewell and Hegel and Mr. W. R. Smith; a Vindication in a Physico-Mathematical Regard. By J. H. STIRLING, LL.D. 8vo. 6s.

As REGARDS PROTOPLASM. By J. H. STIRLING, LL.D. Second Edition, with Additions, in reference to Mr. Huxley's Second Issue and a new Preface in reply to Mr. Huxley in 'Yeast.' 8vo. price 2s.

CAUSAILTY; or, the Philosophy of Law Investigated. By the Rev. GEORGE JAMIESON, B.D. of Old Machar. Second Edition, greatly enlarged. 8vo. price 12s.

The SENSES and the INTELLECT. By ALEXANDER BAIN, M.D. Professor of Logic in the University of Aberdeen. Third Edition. 8vo. 15s.

MENTAL and MORAL SCIENCE: a Compendium of Psychology and Ethics. By the same Author. Third Edition. Crown 8vo. 10s. 6d. Or separately: PART I. *Mental Science*, price 6s. 6d.; PART II. *Moral Science*, price 4s. 6d.

LOGIC, DEDUCTIVE and INDUCTIVE. By the same Author. In TWO PARTS, crown 8vo. 10s. 6d. Each Part may be had separately:— PART I. *Deduction*, 4s. PART II. *Induction*, 6s. 6d.

TIME and SPACE; a Metaphysical Essay. By SHADWORTH H. HODGSON. (This work covers the whole ground of Speculative Philosophy.) 8vo. price 16s.

The THEORY of PRACTICE; an ETHICAL ENQUIRY. By the same Author. (This work, in conjunction with the foregoing, completes a system of Philosophy.) 2 vols. 8vo. price 24s.

The PHILOSOPHY of NECESSITY; or, Natural Law as applicable to Mental, Moral, and Social Science. By CHARLES BRAY. Second Edition. 8vo. 9s.

A Manual of Anthropology, or Science of Man, based on Modern Research. By the same Author. Crown 8vo. price 6s.

On Force, its Mental and Moral Correlates. By the same Author. 8vo. 5s.

A TREATISE on HUMAN NATURE; being an Attempt to Introduce the Experimental Method of Reasoning into Moral Subjects. By DAVID HUME. Edited, with Notes, &c. by T. H. GREEN, Fellow, and T. H. GROSE, late Scholar, of Balliol College, Oxford. 2 vols. 8vo. [*In the press.*

ESSAYS MORAL, POLITICAL, and LITERARY. By DAVID HUME. By the same Editors. 2 vols. 8vo.   [*In the press.*

UEBERWEG'S SYSTEM of LOGIC and HISTORY of LOGICAL
DOCTRINES. Translated, with Notes and Appendices, by T. M. LINDSAY,
M.A. F.R.S.E. 8vo. price 16s.

A BUDGET of PARADOXES. By AUGUSTUS DE MORGAN, F.R.A.S.
and C.P.S. Reprinted, with the Author's Additions, from the *Athenæum.*
8vo. price 15s.

Astronomy, Meteorology, Popular Geography, &c.

OUTLINES of ASTRONOMY. By Sir J. F. W. HERSCHEL, Bart.
M.A. Eleventh Edition, with 9 Plates and numerous Diagrams. Square
crown 8vo. price 12s.

ESSAYS on ASTRONOMY: a Series of Papers on Planets and Meteors,
the Sun and sun-surrounding Space, Stars and Star Cloudlets; and a Disser-
tation on the approaching Transit of Venus: preceded by a Sketch of the
Life and Work of Sir John Herschel. By RICHARD A. PROCTOR, B.A. Hon.
Sec. R.A.S. With 10 Plates and 24 Woodcuts. 8vo. price 12s.

The SUN; RULER, LIGHT, FIRE, and LIFE of the PLANETARY
SYSTEM. By RICHARD A. PROCTOR, B.A. F.R.A.S. Second Edition
with 10 Plates (7 coloured) and 107 Figures on Wood. Crown 8vo. 14s.

OTHER WORLDS THAN OURS; the Plurality of Worlds Studied
under the Light of Recent Scientific Researches. By the same Author.
Second Edition, with 14 Illustrations. Crown 8vo. 10s. 6d.

THE ORBS AROUND US; a Series of Familiar Essays on the Moon
and Planets, Meteors and Comets, the Sun and Coloured Pairs of Stars.
By the same Author. Crown 8vo. price 7s. 6d.

THE STAR DEPTHS; or, Other Suns than Ours; a Treatise on
Stars, Star-Systems, and Star-Cloudlets. By the same Author. Crown 8vo
with numerous Illustrations. [Nearly ready.

SATURN and its SYSTEM. By the same Author. 8vo. with 14 Plates, 14s.

SCHELLEN'S SPECTRUM ANALYSIS, in its application to Terres-
trial Substances and the Physical Constitution of the Heavenly Bodies.
Translated by JANE and C. LASSELL: edited, with Notes, by W. HUGGINS,
LL.D. F.R.S. With 13 Plates (6 coloured) and 223 Woodcuts. 8vo. price 28s.

A NEW STAR ATLAS, for the Library, the School, and the Observatory,
in Twelve Circular Maps (with Two Index Plates). Intended as a Companion to 'Webb's Celestial Objects for Common Telescopes.' With a
Letterpress Introduction on the Study of the Stars, illustrated by 9 Diagrams. By RICHARD A. PROCTOR, B.A. Hon. Sec. R.A.S. Crown 8vo. 5s.

CELESTIAL OBJECTS for COMMON TELESCOPES. By the Rev.
T. W. WEBB, M.A. F.R.A.S. New Edition, revised, with a large Map of
the Moon, and several Woodcuts. Crown 8vo. price 7s. 6d.

AIR and RAIN: the Beginnings of a Chemical Climatology. By
ROBERT ANGUS SMITH, Ph.D. F.R.S. F.C.S. Government Inspector of
Alkali Works. With 8 Illustrations. 8vo. price 24s.

NAUTICAL SURVEYING, an INTRODUCTION to the PRACTICAL
and THEORETICAL STUDY of. By JOHN KNOX LAUGHTON, M.A.
F.R.A.S. Small 8vo. price 6s.

MAGNETISM and DEVIATION of the COMPASS. For the Use of
Students in Navigation and Science Schools. By JOHN MERRIFIELD, LL.D.
F.R.A.S. 18mo. price 1s. 6d.

DOVE'S LAW of STORMS, considered in connexion with the Ordinary Movements of the Atmosphere. Translated by R. H. Scott, M.A. T.C.D. 8vo. 10s. 6d.

A GENERAL DICTIONARY of GEOGRAPHY, Descriptive, Physical, Statistical, and Historical: forming a complete Gazetteer of the World. By A. Keith Johnston, LL.D. F.R.G.S. New Edition, thoroughly revised.
[In the press.

A MANUAL of GEOGRAPHY, Physical, Industrial, and Political. By W. Hughes, F.R.G.S. With 6 Maps. Fcp. 7s. 6d.

MAUNDER'S TREASURY of GEOGRAPHY, Physical, Historical, Descriptive, and Political. Edited by W. Hughes, F.R.G.S. Revised Edition, with 7 Maps and 16 Plates. Fcp. 6s. cloth, or 10s. bound in calf.

The PUBLIC SCHOOLS ATLAS of MODERN GEOGRAPHY. In 31 Maps, exhibiting clearly the more important Physical Features of the Countries delineated, and Noting all the Chief Places of Historical, Commercial, or Social Interest. Edited, with an Introduction, by the Rev. G. Butler, M.A. Imp. 4to. price 3s. 6d. sewed, or 5s. cloth.

Natural History and Popular Science.

TEXT-BOOKS of SCIENCE, MECHANICAL and PHYSICAL.
Edited by T. M. Goodeve, M.A. and C. W. Merrifield, F.R.S.

1. Goodeve's Mechanism, 3s. 6d.
2. Bloxam's Metals, 3s. 6d.
3. Miller's Inorganic Chemistry, 3s. 6d.
4. Griffin's Algebra and Trigonometry, 3s. 6d.
   Notes and Solutions to Algebra and Trigonometry, 3s. 6d.
5. Watson's Plane and Solid Geometry, 3s. 6d.
6. Maxwell's Theory of Heat, 3s. 6d.
7. Merrifield's Technical Arithmetic and Mensuration, 3s. 6d.
   Hunter's Key to Merrifield's Arithmetic and Mensuration, 3s. 6d.
8. Anderson's Strength of Materials and Structures, 3s. 6d.
9. Jenkin's Electricity and Magnetism, 3s. 6d.

ELEMENTARY TREATISE on PHYSICS, Experimental and Applied. Translated and edited from Ganot's Éléments de Physique (with the Author's sanction) by E. Atkinson, Ph.D. F.C.S. New Edition, revised and enlarged; with a Coloured Plate and 726 Woodcuts. Post 8vo. 15s.

NATURAL PHILOSOPHY for GENERAL READERS and YOUNG PERSONS; being a Course of Physics divested of Mathematical Formulæ, expressed in the language of daily life. Translated from Ganot's Cours de Physique, with the Author's sanction, by E. Atkinson, Ph.D. F.C.S Crown 8vo. with 404 Woodcuts, price 7s. 6d.

Mrs. MARCET'S CONVERSATIONS on NATURAL PHILOSOPHY. Revised by the Author's Son, and augmented by Conversations on Spectrum Analysis and Solar Chemistry. With 36 Plates. Crown 8vo. price 7s. 6d.

SOUND: a Course of Eight Lectures delivered at the Royal Institution of Great Britain. By John Tyndall, LL.D. F.R.S. New Edition, crown 8vo. with Portrait of M. Chladni and 169 Woodcuts, price 9s.

HEAT a MODE of MOTION. By Professor John Tyndall, LL.D. F.R.S. Fourth Edition. Crown 8vo. with Woodcuts, 10s. 6d.

CONTRIBUTIONS to MOLECULAR PHYSICS in the DOMAIN of RADIANT HEAT; a Series of Memoirs published in the Philosophical Transactions and Philosophical Magazine. By JOHN TYNDALL, LL.D. F.R.S. With 2 Plates and 31 Woodcuts. 8vo. price 16s.

RESEARCHES on DIAMAGNETISM and MAGNE-CRYSTALLIC ACTION; including the Question of Diamagnetic Polarity. By the same Author. With 6 Plates and many Woodcuts. 8vo. price 14s.

NOTES of a COURSE of SEVEN LECTURES on ELECTRICAL PHENOMENA and THEORIES, delivered at the Royal Institution, A.D. 1870. By JOHN TYNDALL, LL.D. Crown 8vo. 1s. sewed, or 1s. 6d. cloth.

NOTES of a COURSE of NINE LECTURES on LIGHT delivered at the Royal Institution, A.D. 1869. By the same Author. Crown 8vo. price 1s. sewed, or 1s. 6d. cloth.

FRAGMENTS of SCIENCE. By JOHN TYNDALL, LL.D. F.R.S. Third Edition. 8vo. price 14s.

LIGHT SCIENCE for LEISURE HOURS; a Series of Familiar Essays on Scientific Subjects, Natural Phenomena, &c. By R. A. PROCTOR, B.A. F.R.A.S. Second Edition, revised. Crown 8vo. price 7s. 6d.

LIGHT: Its Influence on Life and Health. By FORBES WINSLOW, M.D. D.C.L. Oxon. (Hon.). Fcp. 8vo. 6s.

The CORRELATION of PHYSICAL FORCES. By W. R. GROVE, Q.C. V.P.R.S. Fifth Edition, revised, and followed by a Discourse on Continuity. 8vo. 10s. 6d. The *Discourse on Continuity*, separately, 2s. 6d.

Professor OWEN'S LECTURES on the COMPARATIVE ANATOMY and Physiology of the Invertebrate Animals. Second Edition, with 235 Woodcuts. 8vo. 21s.

The COMPARATIVE ANATOMY and PHYSIOLOGY of the VERTE- brate Animals. By RICHARD OWEN, F.R.S. D.C.L. With 1,472 Woodcuts. 3 vols. 8vo. £3 13s. 6d.

The ANCIENT STONE IMPLEMENTS, WEAPONS, and ORNA- MENTS of GREAT BRITAIN. By JOHN EVANS, F.R.S. F.S.A. With 2 Plates and 476 Woodcuts. 8vo. price 28s.

The ORIGIN of CIVILISATION and the PRIMITIVE CONDITION of MAN: Mental and Social Condition of Savages. By Sir JOHN LUBBOCK, Bart. M.P. F.R.S. Second Edition, with 25 Woodcuts. 8vo. price 16s.

The PRIMITIVE INHABITANTS of SCANDINAVIA: containing a Description of the Implements, Dwellings, Tombs, and Mode of Living of the Savages in the North of Europe during the Stone Age. By SVEN NILSSON. With 16 Plates of Figures and 3 Woodcuts. 8vo. 18s.

MANKIND, their ORIGIN and DESTINY. By an M.A. of Balliol College, Oxford. Containing a New Translation of the First Three Chapters of Genesis; a Critical Examination of the First Two Gospels; an Explanation of the Apocalypse; and the Origin and Secret Meaning of the Mythological and Mystical Teaching of the Ancients. With 31 Illustrations. 8vo. price 31s. 6d.

BIBLE ANIMALS; being a Description of every Living Creature mentioned in the Scriptures, from the Ape to the Coral. By the Rev. J. G. WOOD, M.A. F.L.S. With about 100 Vignettes on Wood. 8vo. 21s.

HOMES WITHOUT HANDS; a Description of the Habitations of Animals, classed according to their Principle of Construction. By the Rev. J. G. WOOD, M.A. F.L.S. With about 140 Vignettes on Wood. 8vo. 21s.

INSECTS AT HOME; a Popular Account of British Insects, their Structure, Habits, and Transformations. By the Rev. J. G. WOOD, M.A. F.L.S. With upwards of 700 Illustrations engraved on Wood (1 coloured and 21 full size of page). 8vo. price 21s.

INSECTS ABROAD; being a Popular Account of Foreign Insects, their Structure, Habits, and Transformations. By J. G. WOOD, M.A. F.L.S. Author of 'Homes without Hands,' &c. In One Volume, printed and illustrated uniformly with 'Insects at Home,' to which it will form a Sequel and Companion. [In the press.

STRANGE DWELLINGS; a description of the Habitations of Animals, abridged from 'Homes without Hands.' By the Rev. J. G. WOOD, M.A. F.L.S. With about 60 Woodcut Illustrations. Crown 8vo. price 7s. 6d.

A FAMILIAR HISTORY of BIRDS. By E. STANLEY, D.D. F.R.S. late Lord Bishop of Norwich. Seventh Edition, with Woodcuts. Fcp. 3s. 6d.

The **HARMONIES of NATURE and UNITY of CREATION.** By Dr. GEORGE HARTWIG. 8vo. with numerous Illustrations, 18s.

The **SEA and its LIVING WONDERS.** By the same Author. Third (English) Edition. 8vo. with many Illustrations, 21s.

The **SUBTERRANEAN WORLD.** By Dr. GEORGE HARTWIG. With 6 Maps and about 80 Woodcuts, including 8 full size of page. 8vo. price 21s.

The **TROPICAL WORLD;** a Popular Scientific Account of the Natural History of the Equatorial Regions. By Dr. GEORGE HARTWIG. New Edition, with about 200 Illustrations. 8vo. price 10s. 6d.

The **POLAR WORLD;** a Popular Description of Man and Nature in the Arctic and Antarctic Regions of the Globe. By Dr. GEORGE HARTWIG. With 8 Chromoxylographs, 3 Maps, and 85 Woodcuts. 8vo. 21s.

KIRBY and SPENCE'S INTRODUCTION to ENTOMOLOGY, or Elements of the Natural History of Insects. 7th Edition. Crown 8vo. 5s.

MAUNDER'S TREASURY of NATURAL HISTORY, or Popular Dictionary of Zoology. Revised and corrected by T. S. COBBOLD, M.D. Fcp. with 900 Woodcuts, 6s. cloth, or 10s. bound in calf.

The **TREASURY of BOTANY,** or Popular Dictionary of the Vegetable Kingdom; including a Glossary of Botanical Terms. Edited by J. LINDLEY, F.R.S. and T. MOORE, F.L.S. assisted by eminent Contributors. With 274 Woodcuts and 20 Steel Plates. Two Parts, fcp. 12s. cloth, or 20s. calf.

HANDBOOK of HARDY TREES, SHRUBS, and HERBACEOUS PLANTS, containing Descriptions, Native Countries, &c. of a Selection of the Best Species in Cultivation; together with Cultural Details, Comparative Hardiness, Suitability for Particular Positions, &c. By W. B. HEMSLEY, formerly Assistant at the Herbarium of the Royal Gardens, Kew. Based on DECAISNE and NAUDIN'S *Manuel de l'Amateur des Jardins,* and including the 264 Original Woodcuts. Medium 8vo. 21s.

A GENERAL SYSTEM of DESCRIPTIVE and ANALYTICAL BOTANY. I. Organography, Anatomy, and Physiology of Plants. II. Iconography, or the Description and History of Natural Families. Translated from the French of E. LE MAOUT, M.D. and J. DECAISNE, Member of the Institute, by Mrs. HOOKER. Edited, and arranged according to the Botanical System adopted in the Universities and Schools of Great Britain, by J. D. HOOKER, M.D. &c. Director of the Royal Botanic Gardens, Kew. With 5,500 Woodcuts from Designs by N. Stenheil and A. Riocreux. Medium 8vo. price 52s. 6d.

An EXPOSITION of FALLACIES in the HYPOTHESIS of Mr.
DARWIN. By C. R. BREE, M.D. F.Z.S. Author of 'Birds of Europe not
observed in the British Isles' &c. With 36 Woodcuts. Crown 8vo. price 14s.

The ELEMENTS of BOTANY for FAMILIES and SCHOOLS.
Tenth Edition, revised by THOMAS MOORE, F.L.S. Fcp. with 154 Woodcuts, 2s. 6d.

The ROSE AMATEUR'S GUIDE. By THOMAS RIVERS. Twelfth
Edition. Fcp. 4s.

LOUDON'S ENCYCLOPÆDIA of PLANTS; comprising the Specific
Character, Description, Culture, History, &c. of all the Plants found in
Great Britain. With upwards of 12,000 Woodcuts. 8vo. 42s.

MAUNDER'S SCIENTIFIC and LITERARY TREASURY. New
Edition, thoroughly revised and in great part re-written, with above 1,000
new Articles, by J. Y. JOHNSON, Corr. M.Z.S. Fcp. 6s. cloth, or 10s. calf.

A DICTIONARY of SCIENCE, LITERATURE, and ART. Fourth
Edition, re-edited by W. T. BRANDE (the original Author), and GEORGE W.
COX, M.A. assisted by contributors of eminent Scientific and Literary
Acquirements. 3 vols. medium 8vo. price 63s. cloth.

Chemistry, Medicine, Surgery, and the Allied Sciences.

A DICTIONARY of CHEMISTRY and the Allied Branches of other
Sciences. By HENRY WATTS, F.R.S. assisted by eminent Contributors
Complete in 5 vols. medium 8vo. £7 3s.

Supplement; bringing the Record of Chemical Discovery down to
the end of the year 1869; including also several Additions to, and Corrections
of, former results which have appeared in 1870 and 1871. By HENRY WATTS,
B.A. F.R.S. F.C.S. Assisted by eminent Scientific and Practical Chemists,
Contributors to the Original Work. 8vo. price 31s. 6d.

ELEMENTS of CHEMISTRY, Theoretical and Practical. By W. ALLEN
MILLER, M.D. late Prof. of Chemistry, King's Coll. London. New
Edition. 3 vols. 8vo. £3. PART I. CHEMICAL PHYSICS, 15s. · PART II.
INORGANIC CHEMISTRY, 21s. PART III. ORGANIC CHEMISTRY, 24s.

A Course of Practical Chemistry, for the use of Medical Students.
By W. ODLING, F.R.S. New Edition, with 70 Woodcuts. Crown 8vo. 7s. 6d.

A MANUAL of CHEMICAL PHYSIOLOGY, including its Points of
Contact with Pathology. By J. L. W. THUDICHUM, M.D. With Woodcuts.
8vo. price 7s. 6d.

SELECT METHODS in CHEMICAL ANALYSIS, chiefly INOR-
GANIC. By WILLIAM CROOKES, F.R.S. With 22 Woodcuts. Crown 8vo.
price 12s. 6d.

CHEMICAL NOTES for the LECTURE ROOM. By THOMAS WOOD,
F.C.S. 2 vols. crown 8vo. I. on Heat &c. price 5s. II. on the Metals, 5s.

The HANDBOOK for MIDWIVES. By HENRY FLY SMITH, B.A.
M.B. Oxon. M.R.C.S. late Assistant-surgeon at the Hospital for Women,
Soho-square. With 41 Woodcuts. Crown 8vo. 5s.

The DIAGNOSIS, PATHOLOGY, and TREATMENT of DISEASES
of Women; including the Diagnosis of Pregnancy. By GRAILY HEWITT,
M.D. Third Edition, partly re-written; with several additional Illustrations. 8vo. price 24s.

On **SOME DISORDERS** of the **NERVOUS SYSTEM** in **CHILD-HOOD**; being the Lumleian Lectures delivered before the Royal College of Physicians in March 1871. By CHARLES WEST, M.D. Crown 8vo. price 5s.

LECTURES on the **DISEASES** of **INFANCY** and **CHILDHOOD**. By CHARLES WEST, M.D. &c. Fifth Edition, revised and enlarged. 8vo. 16s.

The **SCIENCE** and **ART** of **SURGERY**; being a Treatise on Surgical Injuries, Diseases and Operations. By JOHN ERIC ERICHSEN, Senior Surgeon to University College Hospital, and Holme Professor of Clinical Surgery in University College. London A new Edition, being the Sixth, revised and enlarged; with 712 Woodcuts. 2 vols. 8vo. price 32s.

A **SYSTEM** of **SURGERY**, Theoretical and Practical. In Treatises by Various Authors. Edited by T. HOLMES, M.A. &c. Surgeon and Lecturer on Surgery at St. George's Hospital. Second Edition, thoroughly revised, with numerous Illustrations. 5 vols. 8vo. £5 5s.

The **SURGICAL TREATMENT** of **CHILDREN'S DISEASES**. By T. HOLMES. M.A. &c. late Surgeon to the Hospital for Sick Children. Second Edition, with 9 Plates and 112 Woodcuts. 8vo. 21s.

LECTURES on the **PRINCIPLES** and **PRACTICE** of **PHYSIC**. By Sir THOMAS WATSON, Bart. M.D. Fifth Edition, thoroughly revised. 2 vols. 8vo. price 36s.

LECTURES on **SURGICAL PATHOLOGY**. By Sir JAMES PAGET, Bart. F.R.S. Third Edition, revised and re-edited by the Author and Professor W. TURNER, M.B. 8vo. with 131 Woodcuts, 21s.

COOPER'S DICTIONARY of **PRACTICAL SURGERY** and Encyclopædia of Surgical Science. New Edition, brought down to the present time. By S. A. LANE, Surgeon to St. Mary's Hospital, assisted by various Eminent Surgeons. 2 vols. 8vo. price 25s. each.

On **CHRONIC BRONCHITIS**, especially as connected with **GOUT**, **EMPHYSEMA**, and DISEASES of the **HEART**. By E. HEADLAM GREENHOW, M.D. F.R.C.P. &c. 8vo. 7s. 6d.

The **CLIMATE** of the **SOUTH** of **FRANCE** as SUITED to INVALIDS; with Notices of Mediterranean and other Winter Stations. By C. T. WILLIAMS, M.A. M.D. Oxon. Second Edition. Crown 8vo. 6s.

PULMONARY CONSUMPTION; its Nature, Varieties, and Treatment: with an Analysis of One Thousand Cases to exemplify its Duration. By C. J. B. WILLIAMS, M.D. F.R.S. and C. T. WILLIAMS, M.A. M.D. Oxon. Post 8vo. price 10s. 6d.

CLINICAL LECTURES on **DISEASES** of the **LIVER, JAUNDICE**, and ABDOMINAL DROPSY. By CHARLES MURCHISON, M.D. Post 8vo. with 25 Woodcuts, 10s. 6d.

A **TREATISE** on the **CONTINUED FEVERS** of **GREAT BRITAIN**. By CHARLES MURCHISON, M.D. New Edition, revised. [Nearly ready.

QUAIN'S ELEMENTS of **ANATOMY**. Seventh Edition [1867], edited by W. SHARPEY. M.D. F.R.S. ALLEN THOMPSON. M.D. F.R.S. and J. CLELAND. M.D. With upwards of 800 Engravings on Wood. 2 vols. 8vo. price 31s. 6d.

ANATOMY, DESCRIPTIVE and **SURGICAL**. By HENRY GRAY, F.R.S. With about 400 Woodcuts from Dissections. Sixth Edition, by T. HOLMES, M.A. Cantab. with a new Introduction. Royal 8vo. 28s.

OUTLINES of **PHYSIOLOGY**, Human and Comparative. By JOHN MARSHALL, F.R.C.S. Surgeon to the University College Hospital. 2 vols. crown 8vo. with 122 Woodcuts, 32s.

PHYSIOLOGICAL ANATOMY and PHYSIOLOGY of MAN. By the late R. B. TODD, M.D. F.R.S. and W. BOWMAN, F.R.S. of King's College. With numerous Illustrations. VOL. II. 8vo. 25s.
VOL. I. New Edition by Dr. LIONEL S. BEALE, F.R.S. In course of publication, with many Illustrations. PARTS I. and II. price 7s. 6d. each.

COPLAND'S DICTIONARY of PRACTICAL MEDICINE, abridged from the larger work and throughout brought down to the present State of Medical Science. 8vo. 36s.

On the **MANUFACTURE of BEET-ROOT SUGAR in ENGLAND** and IRELAND. By WILLIAM CROOKES, F.R.S. Crown 8vo. with 11 Woodcuts, 8s. 6d.

DR. PEREIRA'S ELEMENTS of MATERIA MEDICA and THERA-PEUTICS, abridged and adapted for the use of Medical and Pharmaceutical Practitioners and Students; and comprising all the Medicines of the British Pharmacopœia, with such others as are frequently ordered in Prescriptions or required by the Physician. Edited by Professor BENTLEY, F.L.S. &c. and by Dr. REDWOOD, F.C.S. &c. With 125 Woodcut Illustrations. 8vo. price 25s.

The **ESSENTIALS of MATERIA MEDICA and THERAPEUTICS.** By ALFRED BARING GARROD, M.D. F.R.S. &c. Physician to King's College Hospital. Third Edition. Sixth Impression, brought up to 1870. Crown 8vo. price 12s. 6d.

The Fine Arts, and *Illustrated Editions.*

GROTESQUE ANIMALS, invented, described, and portrayed by E. W. COOKE, R.A. F.R.S. F.G.S. F.Z.S. &c. In Twenty-four Plates, with Elucidatory Comments. Royal 4to. 21s.

IN FAIRYLAND; Pictures from the Elf-World. By RICHARD DOYLE. With a Poem by W. ALLINGHAM. With Sixteen Plates, containing Thirty-six Designs printed in Colours. Folio, 31s. 6d.

HALF-HOUR LECTURES on the HISTORY and PRACTICE of the Fine and Ornamental Arts. By WILLIAM B. SCOTT. New Edition, revised by the Author; with 50 Woodcuts. Crown 8vo. 8s. 6d.

ALBERT DURER, HIS LIFE and WORKS; including Auto-biographical Papers and Complete Catalogues. By WILLIAM B. SCOTT. With Six Etchings by the Author, and other Illustrations. 8vo. 16s.

The **CHORALE BOOK for ENGLAND:** the Hymns translated by Miss C. WINKWORTH; the Tunes arranged by Prof. W. S. BENNETT and OTTO GOLDSCHMIDT. Fcp. 4to. 12s. 6d.

The **NEW TESTAMENT,** illustrated with Wood Engravings after the Early Masters, chiefly of the Italian School. Crown 4to. 63s. cloth, gilt top; or £5. 5s. elegantly bound in morocco.

LYRA GERMANICA; the Christian Year. Translated by CATHERINE WINKWORTH; with 125 Illustrations on Wood drawn by J. LEIGHTON, F.S.A. 4to. 21s.

LYRA GERMANICA; the Christian Life. Translated by CATHERINE WINKWORTH; with about 200 Woodcut Illustrations by J. LEIGHTON, F.S.A. and other Artists. 4to. 21s.

B

The **LIFE of MAN SYMBOLISED** by the **MONTHS** of the **YEAR.**
Text selected by R. PIGOT; Illustrations on Wood from Original Designs by
J. LEIGHTON, F.S.A. 4to. 42s.

CATS' and FARLIE'S MORAL EMBLEMS; with Aphorisms, Adages,
and Proverbs of all Nations. 121 Illustrations on Wood by J. LEIGHTON
F.S.A. Text selected by R. PIGOT. Imperial 8vo. 31s. 6d.

SACRED and LEGENDARY ART. By Mrs. JAMESON.

Legends of the Saints and Martyrs. New Edition, with 19
Etchings and 187 Woodcuts. 2 vols. square crown 8vo. 31s. 6d.

Legends of the Monastic Orders. New Edition, with 11 Etchings
and 88 Woodcuts. 1 vol. square crown 8vo. 21s.

Legends of the Madonna. New Edition, with 27 Etchings and
165 Woodcuts. 1 vol. square crown 8vo. 21s.

The History of Our Lord, with that of his Types and Precursors.
Completed by Lady EASTLAKE. Revised Edition, with 31 Etchings and
281 Woodcuts. 2 vols. square crown 8vo. 42s.

The Useful Arts, Manufactures, &c.

HISTORY of the GOTHIC REVIVAL; an Attempt to shew how far
the taste for Mediæval Architecture was retained in England during the
last two centuries, and has been re-developed in the present. By C. L. EAST-
LAKE, Architect. With 48 Illustrations (36 full size of page). Imperial 8vo.
price 31s. 6d.

GWILT'S ENCYCLOPÆDIA of ARCHITECTURE, with above 1,600
Engravings on Wood. Fifth Edition, revised and enlarged by WYATT
PAPWORTH. 8vo. 52s. 6d.

A MANUAL of ARCHITECTURE: being a Concise History and
Explanation of the principal Styles of European Architecture, Ancient,
Mediæval, and Renaissance; with a Glossary of Technical Terms. By
THOMAS MITCHELL. Crown 8vo. with 150 Woodcuts, 10s. 6d.

HINTS on HOUSEHOLD TASTE in FURNITURE, UPHOLSTERY,
and other Details. By CHARLES L. EASTLAKE, Architect. New Edition
with about 90 Illustrations. Square crown 8vo. 14s.

PRINCIPLES of MECHANISM, designed for the Use of Students in
the Universities, and for Engineering Students generally. By R
WILLIS, M.A. F.R.S. &c. Jacksonian Professor in the University of Cambridge. Second Edition, enlarged; with 374 Woodcuts. 8vo. 18s.

GEOMETRIC TURNING; comprising a Description of the new Geo
metric Chuck constructed by Mr. Plant of Birmingham, with directions for
its use, and a series of Patterns cut by it, with Explanations of the mode of
producing them, and an account of a New Process of Deep Cutting and
of Graving on Copper. By H. S. SAVORY. With numerous Woodcuts.
8vo. 21s.

LATHES and TURNING, Simple, Mechanical, and ORNAMENTAL.
By W. HENRY NORTHCOTT. With about 240 Illustrations on Steel and
Wood. 8vo. 18s.

PERSPECTIVE; or, the Art of Drawing what one Sees. Explained
and adapted to the use of those Sketching from Nature. By Lieut. W. H.
COLLINS, R.E. F.R.A.S. With 37 Woodcuts. Crown 8vo. price 5s.

URE'S DICTIONARY of ARTS, MANUFACTURES, and MINES.
Sixth Edition, chiefly rewritten and greatly enlarged by ROBERT HUNT, F.R.S. assisted by numerous Contributors eminent in Science and the Arts, and familiar with Manufactures. With above 2,000 Woodcuts. 3 vols. medium 8vo. price £4. 14s. 6d.

HANDBOOK of PRACTICAL TELEGRAPHY. By R. S. CULLEY, Memb. Inst. C.E. Engineer-in-Chief of Telegraphs to the Post Office. Fifth Edition, with 118 Woodcuts and 9 Plates. 8vo. price 14s.

ENCYCLOPÆDIA of CIVIL ENGINEERING, Historical, Theoretical, and Practical. By E. CRESY, C.E. With above 3,000 Woodcuts. 8vo. 42s.

The STRAINS in TRUSSES Computed by means of Diagrams ; with 20 Examples drawn to Scale. By F. A. RANKEN, M.A. C.E. Lecturer at the Hartley Institution, Southampton. With 35 Diagrams. Square crown 8vo. price 6s. 6d.

TREATISE on MILLS and MILLWORK. By Sir W. FAIRBAIRN, Bart. F.R.S. New Edition, with 18 Plates and 322 Woodcuts. 2 vols. 8vo. 32s.

USEFUL INFORMATION for ENGINEERS. By the same Author. FIRST, SECOND, and THIRD SERIES, with many Plates and Woodcuts. 3 vols. crown 8vo. 10s. 6d. each.

The APPLICATION of CAST and WROUGHT IRON to Building Purposes. By Sir W. FAIRBAIRN, Bart. F.R.S. Fourth Edition, enlarged; with 6 Plates and 118 Woodcuts. 8vo. price 16s.

A TREATISE on the STEAM ENGINE, in its various Applications to Mines, Mills, Steam Navigation, Railways and Agriculture. By J. BOURNE, C.E. Eighth Edition ; with Portrait, 37 Plates, and 546 Woodcuts. 4to. 42s.

CATECHISM of the STEAM ENGINE, in its various Applications to Mines, Mills, Steam Navigation, Railways, and Agriculture. By the same Author. With 89 Woodcuts. Fcp. 6s.

HANDBOOK of the STEAM ENGINE. By the same Author, forming a KEY to the Catechism of the Steam Engine, with 67 Woodcuts. Fcp. 9s.

BOURNE'S RECENT IMPROVEMENTS in the STEAM ENGINE in its various applications to Mines, Mills, Steam Navigation, Railways, and Agriculture. Being a Supplement to the Author's 'Catechism of the Steam Engine.' By JOHN BOURNE, C.E. New Edition, including many New Examples; with 124 Woodcuts. Fcp. 8vo. 6s.

PRACTICAL TREATISE on METALLURGY, adapted from the last German Edition of Professor KERL's *Metallurgy* by W. CROOKES, F.R.S. &c. and E. RÖHRIG, Ph.D. M E With 625 Woodcuts. 3 vols. 8vo. price £4. 19s.

MITCHELL'S MANUAL of PRACTICAL ASSAYING. A New Edition, being the Fourth, thoroughly revised, with recent Discoveries incorporated, by W. CROOKES, F. R.S. With numerous Woodcuts.
[*Nearly ready.*

LOUDON'S ENCYCLOPÆDIA of AGRICULTURE: comprising the Laying-out, Improvement, and Management of Landed Property, and the Cultivation and Economy of the Productions of Agriculture. With 1,100 Woodcuts. 8vo. 21s.

Loudon's Encyclopædia of Gardening: comprising the Theory and Practice of Horticulture, Floriculture, Arboriculture, and Landscape Gardening. With 1,000 Woodcuts. 8vo. 21s.

BAYLDON'S ART of VALUING RENTS and TILLAGES, and Claims of Tenants upon Quitting Farms, both at Michaelmas and Lady-Day. Eighth Edition, revised by J. C. MORTON. 8vo. 10s. 6d.

B 2

Religious and Moral Works.

EIGHT ESSAYS on ECCLESIASTICAL REFORM, by Various Writers ; together with a Preface and Analysis of the Essays. Edited by the Rev. ORBY SHIPLEY, M.A. Crown 8vo. 10s. 6d.

The SPEAKER'S BIBLE COMMENTARY, by Bishops and other Clergy of the Anglican Church, critically examined by the Right Rev. J. W. COLENSO, D.D. Bishop of Natal. 8vo. PART I. *Genesis*, 3s. 6d. PART II. *Exodus*, 4s. 6d. PART III. *Leviticus*, 2s. 6d. PART IV. *Numbers*, 3s. 6d. PART V. *Deuteronomy*, price 5s.

The OUTLINES of the CHRISTIAN MINISTRY DELINEATED, and brought to the Test of Reason, Holy Scripture, History, and Experience, with a view to the Reconciliation of Existing Differences concerning it, especially between Presbyterians and Episcopalians. By C. WORDSWORTH, D.C.L. Bishop of St. Andrews. Crown 8vo. price 7s. 6d.

CHRISTIAN COUNSELS, Selected from the Devotional Works of Fénelon, Archbishop of Cambrai. Translated by A. M. JAMES. Crown 8vo. price 5s.

CHRIST the CONSOLER; a Book of Comfort for the Sick. With a Preface by the Right Rev. the Lord Bishop of Carlisle. Small 8vo. price 6s.

AUTHORITY and CONSCIENCE; a Free Debate on the Tendency of Dogmatic Theology and on the Characteristics of Faith. Edited by CONWAY MOREL. Post 8vo. price 7s. 6d.

REASONS of FAITH; or, the ORDER of the Christian Argument Developed and Explained. By the Rev. G. S. DREW, M.A. Second Edition, revised and enlarged. Fcp. 8vo. price 6s.

The TRUE DOCTRINE of the EUCHARIST. By THOMAS S. L. VOGAN, D.D. Canon and Prebendary of Chichester and Rural Dean. 8vo. price 18s.

CHRISTIAN SACERDOTALISM, viewed from a Layman's standpoint or tried by Holy Scripture and the Early Fathers ; with a short Sketch of the State of the Church from the end of the Third to the Reformation in the beginning of the Sixteenth Century. By JOHN JARDINE, M.A. LL.D. 8vo. price 8s. 6d.

SYNONYMS of the OLD TESTAMENT, their BEARING on CHRISTIAN FAITH and PRACTICE. By the Rev. ROBERT BAKER GIRDLESTONE, M.A. 8vo. price 15s.

An INTRODUCTION to the THEOLOGY of the CHURCH of ENGLAND, in an Exposition of the Thirty-nine Articles. By the Rev. T. P. BOULTBEE, LL.D. Fcp. 8vo. price 6s.

FUNDAMENTALS; or, Bases of Belief concerning MAN and GOD: a Handbook of Mental, Moral, and Religious Philosophy. By the Rev. T. GRIFFITH, M.A. 8vo. price 10s. 6d.

PRAYERS for the FAMILY and for PRIVATE USE, selected from the COLLECTION of the late BARON BUNSEN, and Translated by CATHERINE WINKWORTH. Fcp. 8vo. price 3s. 6d.

The STUDENT'S COMPENDIUM of the BOOK of COMMON PRAYER; being Notes Historical and Explanatory of the Liturgy of the Church of England. By the Rev. H. ALLDEN NASH. Fcp. 8vo. price 2s. 6d.

CHURCHES and their CREEDS. By the Rev. Sir PHILIP PERRING, Bart. late Scholar of Trin. Coll. Cambridge, and University Medallist. Crown 8vo. price 10s. 6d.

An EXPOSITION of the 39 ARTICLES, Historical and Doctrinal.
By E. HAROLD BROWNE, D.D. Lord Bishop of Ely. Ninth Edit. 8vo. 16s.

The LIFE and EPISTLES of ST. PAUL. By the Rev. W. J.
CONYBEARE, M.A., and the Very Rev. J. S. HOWSON, D.D. Dean of Chester :—
LIBRARY EDITION, with all the Original Illustrations, Maps, Landscapes
on Steel, Woodcuts, &c. 2 vols. 4to. 48s.
INTERMEDIATE EDITION, with a Selection of Maps, Plates, and Woodcuts.
2 vols. square crown 8vo. 21s.
STUDENT'S EDITION, revised and condensed, with 46 Illustrations and
Maps. 1 vol. crown 8vo. price 9s.

The VOYAGE and SHIPWRECK of ST. PAUL; with Dissertations
on the Life and Writings of St. Luke and the Ships and Navigation of the
Ancients. By JAMES SMITH, F.R.S. Third Edition. Crown 8vo. 10s. 6d.

COMMENTARY on the EPISTLE to the ROMANS. By the Rev.
W. A. O'CONNOR, B.A. Rector of St. Simon and St. Jude, Manchester. Crown
8vo. price 3s. 6d.

The EPISTLE to the HEBREWS; with Analytical Introduction and
Notes. By the Rev. W. A. O'CONNOR, B.A. Crown 8vo. price 4s. 6d.

A CRITICAL and GRAMMATICAL COMMENTARY on ST. PAUL'S
Epistles. By C. J. ELLICOTT, D.D. Lord Bishop of Gloucester & Bristol. 8vo.

Galatians, Fourth Edition, 8s. 6d.

Ephesians, Fourth Edition, 8s. 6d.

Pastoral Epistles, Fourth Edition, 10s. 6d.

Philippians, Colossians, and Philemon, Third Edition, 10s. 6d.

Thessalonians, Third Edition, 7s. 6d.

HISTORICAL LECTURES on the LIFE of OUR LORD JESUS
CHRIST: being the Hulsean Lectures for 1859. By C. J. ELLICOTT, D.D.
Lord Bishop of Gloucester and Bristol. Fifth Edition. 8vo. price 12s.

EVIDENCE of the TRUTH of the CHRISTIAN RELIGION derived
from the Literal Fulfilment of Prophecy. By ALEXANDER KEITH, D.D.
37th Edition, with numerous Plates, in square 8vo. 12s. 6d.; also the 39th
Edition, in post 8vo. with 5 Plates, 6s.

History and Destiny of the World and Church, according to
Scripture. By the same Author. Square 8vo. with 40 Illustrations, 10s.

An INTRODUCTION to the STUDY of the NEW TESTAMENT,
Critical, Exegetical, and Theological. By the Rev. S. DAVIDSON, D.D.
LL.D. 2 vols. 8vo. 30s.

EWALD'S HISTORY of ISRAEL to the DEATH of MOSES. Translated from the German. Edited, with a Preface and an Appendix, by RUSSELL
MARTINEAU, M.A. Second Edition. 2 vols. 8vo. 24s. VOLS. III. and IV.
edited by J. E. CARPENTER, M.A. price 21s.

The HISTORY and LITERATURE of the ISRAELITES, according
to the Old Testament and the Apocrypha. By C. DE ROTHSCHILD and
A. DE ROTHSCHILD. Second Edition, revised. 2 vols. post 8vo. with Two
Maps, price 12s. 6d. Abridged Edition, in 1 vol. fcp. 8vo. price 3s. 6d.

The TREASURY of BIBLE KNOWLEDGE; being a Dictionary of the
Books, Persons, Places, Events, and other matters of which mention is made
in Holy Scripture. By Rev. J. AYRE, M.A. With Maps, 16 Plates, and
numerous Woodcuts. Fcp. 8vo. price 6s. cloth, or 10s. neatly bound in calf.

The **GREEK TESTAMENT**; with Notes, Grammatical and Exegetical. By the Rev. W. WEBSTER, M.A. and the Rev. W. F. WILKINSON, M.A. 2 vols. 8vo. £2 4s.

EVERY-DAY SCRIPTURE DIFFICULTIES explained and illustrated. By J. E. PRESCOTT, M.A. VOL. I. *Matthew* and *Mark*; VOL. II. *Luke* and *John*. 2 vols. 8vo. 9s. each.

The **PENTATEUCH** and **BOOK** of **JOSHUA CRITICALLY EXAMINED**. By the Right Rev. J. W. COLENSO, D.D. Lord Bishop of Natal. People's Edition, in 1 vol. crown 8vo. 6s. PART VI. *the Later Legislation of the Pentateuch.* 8vo. price 24s.

The **FORMATION** of **CHRISTENDOM**. By T. W. ALLIES. PARTS I. and II. 8vo. price 12s. each Part.

ENGLAND and **CHRISTENDOM**. By ARCHBISHOP MANNING, D.D. Post 8vo. price 10s. 6d.

A **VIEW** of the **SCRIPTURE REVELATIONS CONCERNING** a FUTURE STATE. By RICHARD WHATELY, D.D. late Archbishop of Dublin. Ninth Edition. Fcp. 8vo. 5s.

THOUGHTS for the **AGE**. By ELIZABETH M. SEWELL, Author of 'Amy Herbert' &c. New Edition, revised. Fcp. 8vo. price 5s.

Passing Thoughts on Religion. By the same Author. Fcp. 8vo. 3s. 6d.

Self-Examination before Confirmation. By the same Author. 32mo. price 1s. 6d.

Readings for a **Month Preparatory** to **Confirmation**, from Writers of the Early and English Church. By the same Author. Fcp. 4s.

Readings for **Every Day** in **Lent**, compiled from the Writings of Bishop JEREMY TAYLOR. By the same Author. Fcp. 5s.

Preparation for the **Holy Communion**; the Devotions chiefly from the works of JEREMY TAYLOR. By the same Author. 32mo. 3s.

THOUGHTS for the **HOLY WEEK** for Young Persons. By the Author of 'Amy Herbert.' New Edition. Fcp. 8vo. 2s.

PRINCIPLES of **EDUCATION** Drawn from **Nature** and **Revelation**, and applied to Female Education in the Upper Classes. By the Author of 'Amy Herbert.' 2 vols. fcp. 12s. 6d.

LYRA GERMANICA, translated from the German by Miss C. WINKWORTH. FIRST SERIES, Hymns for the Sundays and Chief Festivals. SECOND SERIES, the Christian Life. Fcp. 3s. 6d. each SERIES.

SPIRITUAL SONGS for the **SUNDAYS** and **HOLIDAYS** throughout the Year. By J. S. B. MONSELL, LL.D. Vicar of Egham and Rural Dean. Fourth Edition, Sixth Thousand. Fcp. 4s. 6d.

TRADITIONS and **CUSTOMS** of **CATHEDRALS**. By MACKENZIE E. C. WALCOTT, B.D. F.S.A. Precentor and Prebendary of Chichester. Second Edition, revised and enlarged. Crown 8vo. price 6s.

ENDEAVOURS after the **CHRISTIAN LIFE**: Discourses. By JAMES MARTINEAU. Fourth Edition, carefully revised. Post 8vo. 7s. 6d.

WHATELY'S INTRODUCTORY LESSONS on the **CHRISTIAN** Evidences. 18mo. 6d.

FOUR DISCOURSES of **CHRYSOSTOM**, chiefly on the Parable of the Rich Man and Lazarus. Translated by F. ALLEN, B.A. Crown 8vo. 3s. 6d.

BISHOP JEREMY TAYLOR'S ENTIRE WORKS. With Life by
BISHOP HEBER. Revised and corrected by the Rev. C. P. EDEN. 10 vols.
price £5. 5s.

Travels, Voyages, &c.

RAMBLES, by PATRICIUS WALKER. Reprinted from *Fraser's Magazine*
with a Vignette of the Queen's Bower in the New Forest. Crown 8vo. 10s. 6d.

SLAVE-CATCHING in the INDIAN OCEAN; A Record of Naval
Experiences. By Capt. COLOMB, R.N. 8vo. with Illustrations from Photo-
graphs, &c.   [*Nearly ready.*

UNTRODDEN PEAKS and UNFREQUENTED VALLEYS; A Mid-
summer Ramble among the Dolomites, by AMELIA B. EDWARDS. With a
Map and numerous Illustrations, engraved on Wood by E. WHYMPER.
Medium 8vo.   [*In the Spring.*

SIX MONTHS in CALIFORNIA. By J. G. PLAYER-FROWD. Post
8vo. price 6s.

The JAPANESE in AMERICA. By CHARLES LANMAN, American
Secretary, Japanese Legation, Washington, U.S.A. Post 8vo. price 10s. 6d.

MY WIFE and I in QUEENSLAND; Eight Years' Experience in
the Colony, with some account of Polynesian Labour. By CHARLES H.
EDEN. With Map and Frontispiece. Crown 8vo. price 9s.

LIFE in INDIA; a Series of Sketches showing something of the
Anglo-Indian, the Land he lives in, and the People among whom he lives.
By EDWARD BRADDON. Post 8vo. price 9s.

HOW to SEE NORWAY. By Captain J. R. CAMPBELL. With Map
and 5 Woodcuts. Fcp. 8vo. price 5s.

PAU and the PYRENEES. By Count HENRY RUSSELL, Member of
the Alpine Club, &c. With 2 Maps. Fcp. 8vo. price 5s.

CADORE; or, TITIAN'S COUNTRY. By JOSIAH GILBERT, one of
the Authors of 'The Dolomite Mountains.' With Map, Facsimile, and 40
Illustrations. Imperial 8vo. 31s. 6d.

HOURS of EXERCISE in the ALPS. By JOHN TYNDALL, LL.D.
F.R.S. Third Edition, with 7 Woodcuts by E. WHYMPER. Crown 8vo.
price 12s. 6d.

TRAVELS in the CENTRAL CAUCASUS and BASHAN. Including
Visits to Ararat and Tabreez and Ascents of Kazbek and Elbruz. By
D. W. FRESHFIELD. Square crown 8vo. with Maps, &c. 18s.

MAP of the CHAIN of MONT BLANC, from an actual Survey in
1863—1864. By A. ADAMS-REILLY, F.R.G.S. M.A.C. Published under the
Authority of the Alpine Club. In Chromolithography on extra stout
drawing-paper 36in. × 17in. price 10s. or mounted on canvas in a folding
case, 12s. 6d.

HISTORY of DISCOVERY in our AUSTRALASIAN COLONIES,
Australia, Tasmania, and New Zealand, from the Earliest Date to the
Present Day. By WILLIAM HOWITT. 2 vols. 8vo. with 3 Maps, 20s.

The DOLOMITE MOUNTAINS; Excursions through Tyrol, Carinthia,
Carniola, and Friuli, 1861—1863. By J. GILBERT and G. C. CHURCHILL,
F.R.G.S. With numerous Illustrations. Square crown 8vo. 21s.

GUIDE to the PYRENEES, for the use of Mountaineers. By CHARLES PACKE. 2nd Edition, with Map and Illustrations. Cr. 8vo. 7s. 6d.

The ALPINE GUIDE. By JOHN BALL, M.R.I.A. late President of the Alpine Club. Thoroughly Revised Editions, in Three Volumes, post 8vo. with Maps and other Illustrations:—

GUIDE to the WESTERN ALPS, including Mont Blanc, Monte Rosa, Zermatt, &c. Price 6s. 6d.

GUIDE to the CENTRAL ALPS, including all the Oberland District. Price 7s. 6d.

GUIDE to the EASTERN ALPS, price 10s. 6d.

Introduction on Alpine Travelling in General, and on the Geology of the Alps, price 1s. Each of the Three Volumes or Parts of the *Alpine Guide* may be had with this INTRODUCTION prefixed, price 1s. extra.

VISITS to REMARKABLE PLACES: Old Halls, Battle-Fields, and Stones Illustrative of Striking Passages in English History and Poetry By WILLIAM HOWITT. 2 vols. square crown 8vo. with Woodcuts, 25s.

The RURAL LIFE of ENGLAND. By the same Author. With Woodcuts by Bewick and Williams. Medium 8vo. 12s. 6d.

Works of *Fiction.*

POPULAR ROMANCES of the MIDDLE AGES. By GEORGE W COX, M.A. Author of 'The Mythology of the Aryan Nations' &c. and EUSTACE HINTON JONES. Crown 8vo. price 10s. 6d.

TALES of the TEUTONIC LANDS; a Sequel to 'Popular Romances of the Middle Ages.' By the same Authors. Crown 8vo. 10s. 6d.

The BURGOMASTER'S FAMILY; or, Weal and Woe in a Little World. By CHRISTINE MULLER, Translated from the Dutch by Sir JOHN SHAW LEFEVRE, F.R.S. Crown 8vo. price 6s.

NOVELS and TALES. By the Right Hon. B. DISRAELI, M.P. Cabinet Edition, complete in Ten Volumes, crown 8vo. price 6s. each, as follows:—

LOTHAIR, 6s.	HENRIETTA TEMPLE, 6s.
CONINGSBY, 6s.	CONTARINI FLEMING, &c. 6s.
SYBIL, 6s.	ALROY, IXION, &c. 6s.
TANCRED, 6s.	*The* YOUNG DUKE, &c. 6s.
VENETIA, 6s.	VIVIAN GREY, 6s.

The MODERN NOVELIST'S LIBRARY. Each Work, in crown 8vo. complete in a Single Volume:—

MELVILLE'S GLADIATORS, 2s. boards; 2s. 6d. cloth.
———————— GOOD FOR NOTHING, 2s. boards; 2s. 6d. cloth.
———————— HOLMBY HOUSE, 2s. boards; 2s. 6d. cloth.
———————— INTERPRETER, 2s. boards; 2s. 6d. cloth.
———————— KATE COVENTRY, 2s. boards; 2s. 6d. cloth.
———————— QUEEN'S MARIES, 2s. boards; 2s. 6d. cloth.
———————— DIGBY GRAND, 2s. boards; 2s. 6d. cloth.
———————— GENERAL BOUNCE, 2s boards; 2s. 6d. cloth.
TROLLOPE'S WARDEN, 1s. 6d. boards; 2s. cloth.
———————— BARCHESTER TOWERS, 2s. boards; 2s. 6d. cloth.
BRAMLEY-MOORE'S SIX SISTERS *of the* VALLEYS, 2s. boards; 2s. 6d. cloth.

CABINET EDITION of STORIES and TALES by Miss SEWELL:—

AMY HERBERT, 2s. 6d.
GERTRUDE, 2s. 6d.
The EARL'S DAUGHTER, 2s. 6d.
EXPERIENCE of LIFE, 2s. 6d.
CLEVE HALL, 2s. 6d.

IVORS, 2s. 6d.
KATHARINE ASHTON, 2s. 6d.
MARGARET PERCIVAL, 3s. 6d.
LANETON PARSONAGE, 2s. 6d.
URSULA, 3s. 6d.

WONDERFUL STORIES from NORWAY, SWEDEN, and ICELAND. Adapted and arranged by JULIA GODDARD. With an Introductory Essay by the Rev. G. W. COX, M.A. and Six Woodcuts. Square post 8vo. 6s.

BECKER'S GALLUS; or, Roman Scenes of the Time of Augustus: with Notes and Excursuses. New Edition. Post 8vo. 7s. 6d.

BECKER'S CHARICLES; a Tale illustrative of Private Life among the Ancient Greeks: with Notes and Excursuses. New Edition. Post 8vo. 7s. 6d.

TALES of ANCIENT GREECE. By GEORGE W. COX, M.A. late Scholar of Trin. Coll. Oxon. Crown 8vo. price 6s. 6d.

Poetry and The Drama.

BALLADS and LYRICS of OLD FRANCE; with other Poems. By A. LANG, Fellow of Merton College, Oxford. Square fcp. 8vo. price 5s.

MOORE'S IRISH MELODIES, Maclise's Edition, with 161 Steel Plates from Original Drawings. Super-royal 8vo. 31s. 6d.

Miniature Edition of Moore's Irish Melodies with Maclise's Designs (as above) reduced in Lithography. Imp. 16mo. 10s. 6d.

MOORE'S LALLA ROOKH. Tenniel's Edition, with 68 Wood Engravings from original Drawings and other Illustrations. Fcp. 4to. 21s.

SOUTHEY'S POETICAL WORKS, with the Author's last Corrections and copyright Additions. Library Edition, in 1 vol. medium 8vo. with Portrait and Vignette, 14s.

LAYS of ANCIENT ROME; with Ivry and the Armada. By the Right Hon. LORD MACAULAY. 16mo. 3s. 6d.

Lord Macaulay's Lays of Ancient Rome. With 90 Illustrations on Wood, from the Antique, from Drawings by G. SCHARF. Fcp. 4to. 21s.

Miniature Edition of Lord Macaulay's Lays of Ancient Rome, with the Illustrations (as above) reduced in Lithography. Imp. 16mo. 10s. 6d.

GOLDSMITH'S POETICAL WORKS, with Wood Engravings from Designs by Members of the ETCHING CLUB. Imperial 16mo. 7s. 6d.

The ÆNEID of VIRGIL Translated into English Verse. By JOHN CONINGTON, M.A. New Edition. Crown 8vo. 9s.

The **ODES** and **EPODES** of **HORACE**; a Metrical Translation into English, with Introduction and Commentaries. By Lord LYTTON. With Latin Text. New Edition. Post 8vo. price 10s. 6d.

HORATII OPERA. Library Edition, with Marginal References and English Notes. Edited by the Rev. J. E. YONGE. 8vo. 21s.

BOWDLER'S FAMILY SHAKSPEARE, cheaper Genuine Editions. Medium 8vo. large type, with 36 WOODCUTS, price 14s. Cabinet Edition, with the same ILLUSTRATIONS, 6 vols. fcp. 3s. 6d. each.

POEMS. By JEAN INGELOW. 2 vols. fcp. 8vo. price 10s.

FIRST SERIES, containing 'DIVIDED,'' The STAR's MONUMENT,' &c. Sixteenth Thousand. Fcp. 8vo. price 5s.
SECOND SERIES, ' A STORY of DOOM,' ' GLADYS and her ISLAND,' &c. Fifth Thousand. Fcp. 8vo. price 5s.

POEMS by Jean Ingelow. FIRST SERIES, with nearly 100 Illustrations, engraved on Wood by the Brothers DALZIEL. Fcp. 4to. 21s.

Rural Sports, &c.

ENCYCLOPÆDIA of RURAL SPORTS; a complete Account, Historical, Practical, and Descriptive, of Hunting, Shooting, Fishing, Racing, and all other Rural and Athletic Sports and Pastimes. By D. P. BLAINE. With above 600 Woodcuts (20 from Designs by JOHN LEECH). 8vo. 21s.

The **DEAD SHOT**, or Sportsman's Complete Guide; a Treatise on the Use of the Gun, Dog-breaking, Pigeon-shooting, &c. By MARKSMAN. Revised Edition. Fcp. 8vo. with Plates, 5s.

The **FLY-FISHER'S ENTOMOLOGY.** By ALFRED RONALDS. With coloured Representations of the Natural and Artificial Insect. Sixth Edition; with 20 coloured Plates. 8vo. 14s.

A **BOOK on ANGLING**; a complete Treatise on the Art of Angling in every branch. By FRANCIS FRANCIS. New Edition, with Portrait and 15 other Plates, plain and coloured. Post 8vo. 15s.

WILCOCKS'S SEA-FISHERMAN; comprising the Chief Methods of Hook and Line Fishing in the British and other Seas, a Glance at Nets, and Remarks on Boats and Boating. Second Edition, enlarged; with 80 Woodcuts. Post 8vo. 12s. 6d.

HORSES and STABLES. By Colonel F. FITZWYGRAM, XV. the King's Hussars. With Twenty-four Plates of Illustrations, containing very numerous Figures engraved on Wood. 8vo. 15s.

The **HORSE'S FOOT**, and **HOW to KEEP IT SOUND.** By W. MILES, Esq. Ninth Edition, with Illustrations. Imperial 8vo. 12s. 6d.

A PLAIN TREATISE on HORSE-SHOEING. By the same Author. Sixth Edition. Post 8vo. with Illustrations, 2s. 6d.

STABLES and STABLE-FITTINGS. By the same. Imp. 8vo. with 13 Plates, 15s.

REMARKS on HORSES' TEETH, addressed to Purchasers. By the same. Post 8vo. 1s. 6d.

A TREATISE on HORSE-SHOEING and LAMENESS. By JOSEPH GAMGEE, Veterinary Surgeon, formerly Lecturer on the Principles and Practice of Farriery in the New Veterinary College, Edinburgh. 8vo. with 55 Woodcuts, price 15s.

BLAINE'S VETERINARY ART; a Treatise on the Anatomy, Physiology, and Curative Treatment of the Diseases of the Horse, Neat Cattle and Sheep. Seventh Edition, revised and enlarged by C. STEEL, M.R.C.V.S.L. 8vo. with Plates and Woodcuts, 18s.

The HORSE: with a Treatise on Draught. By WILLIAM YOUATT. New Edition, revised and enlarged. 8vo. with numerous Woodcuts, 12s. 6d.

The DOG. By the same Author. 8vo. with numerous Woodcuts, 6s.

The DOG in HEALTH and DISEASE. By STONEHENGE. With 70 Wood Engravings. Square crown 8vo. 7s. 6d.

The GREYHOUND. By STONEHENGE. Revised Edition, with 24 Portraits of Greyhounds. Square crown 8vo. 10s. 6d.

The SETTER; with Notices of the most Eminent Breeds now Extant, Instructions how to Breed, Rear, and Break; Dog Shows, Field Trials, and General Management, &c. By EDWARD LAVERACK. Crown 4to. with 2 plates, price 7s. 6d.

The OX; his Diseases and their Treatment: with an Essay on Parturition in the Cow. By J. R. DOBSON. Crown 8vo. with Illustrations. 7s. 6d.

Works of Utility and General Information.

The THEORY and PRACTICE of BANKING. By H. D. MACLEOD, M.A. Barrister-at-Law. Second Edition, entirely remodelled. 2 vols. 8vo. 30s.

A DICTIONARY, Practical, Theoretical, and Historical, of Commerce and Commercial Navigation. By J. R. M'CULLOCH. New and thoroughly revised Edition. 8vo. price 63s. cloth, or 70s. half-bd. in russia.

The CABINET LAWYER; a Popular Digest of the Laws of England, Civil, Criminal, and Constitutional: intended for Practical Use and General Information. Twenty-third Edition. Fcp. 8vo. price 7s. 6d.

A PROFITABLE BOOK UPON DOMESTIC LAW; Essays for English Women and Law Students. By PERKINS, Junior, M.A. Barrister-at-Law Post 8vo. 10s. 6d.

BLACKSTONE ECONOMISED, a Compendium of the Laws of England to the Present time; in Four Books, each embracing the Legal Principles and Practical Information contained in the respective volumes of Blackstone, supplemented by Subsequent Statutory Enactments, Important Legal Decisions, &c. By D. M. AIRD, of the Middle Temple. Barrister-at-Law. Post 8vo. 7s. 6d.

A HISTORY and EXPLANATION of the STAMP DUTIES: containing Remarks on the Origin of Stamp Duties; a History of the Stamp Duties from their Commencement to the Present Time; Observations on the Past and Present State of the Stamp Laws; an Explanation of the System and Administration of the Tax; Observations on the Stamp Duties in force in Foreign Countries; and the Stamp Laws at present in force in the United Kingdom. By STEPHEN DOWELL, M.A. Assistant-Solicitor of Inland Revenue. 8vo. 12s. 6d.

PEWTNER'S COMPREHENSIVE SPECIFIER; a Guide to the Practical Specification of every kind of Building-Artificers' Work; with Forms of Building Conditions and Agreements, an Appendix, Foot-Notes, and a copious Index. Edited by W. YOUNG, Architect. Crown 8vo. price 6s.

COLLIERIES and COLLIERS; a Handbook of the Law and Leading Cases relating thereto. By J. C. FOWLER, of the Inner Temple, Barrister. Third Edition. Fcp. 8vo. 7s. 6d.

The MATERNAL MANAGEMENT of CHILDREN in HEALTH and Disease. By THOMAS BULL, M.D. Fcp. 5s.

HINTS to MOTHERS on the MANAGEMENT of their HEALTH during the Period of Pregnancy and in the Lying-in Room. By the late THOMAS BULL, M.D. Fcp. 5s.

HOW to NURSE SICK CHILDREN; containing Directions which may be found of service to all who have charge of the Young. By CHARLES WEST, M.D. Second Edition. Fcp. 8vo. 1s. 6d.

NOTES on LYING-IN INSTITUTIONS; with a Proposal for Organising an Institution for Training Midwives and Midwifery Nurses. By FLORENCE NIGHTINGALE. With 5 Plans. Square crown 8vo. 7s. 6d.

CHESS OPENINGS. By F. W. LONGMAN, Balliol College, Oxford. Fcp. 8vo. 2s. 6d.

THE THEORY OF THE MODERN SCIENTIFIC GAME OF WHIST. By WILLIAM POLE, F.R.S. Mus. Doc. Oxon. Fifth Edition, enlarged. Fcp. 8vo. price 2s. 6d.

A PRACTICAL TREATISE on BREWING; with Formulæ for Public Brewers, and Instructions for Private Families. By W. BLACK. 8vo. 10s. 6d.

MODERN COOKERY for PRIVATE FAMILIES, reduced to a System of Easy Practice in a Series of carefully-tested Receipts. By ELIZA ACTON. Newly revised and enlarged Edition; with 8 Plates of Figures and 150 Woodcuts. Fcp. 6s.

WILLICH'S POPULAR TABLES, for ascertaining, according to the Carlisle Table of Mortality, the value of Lifehold, Leasehold, and Church Property, Renewal Fines, Reversions, &c. Seventh Edition, edited by MONTAGUE MARRIOTT, Barrister-at-Law. Post 8vo. price 10s.

MAUNDER'S TREASURY of KNOWLEDGE and LIBRARY of Reference; comprising an English Dictionary and Grammar, Universal Gazetteer, Classical Dictionary, Chronology, Law Dictionary, a Synopsis of the Peerage, useful Tables &c. Revised Edition. Fcp. 8vo. price 6s.

INDEX

Spottiswoode & Co., Printers, New-street Square, London.

www.ingramcontent.com/pod-product-compliance
Lightning Source LLC
Chambersburg PA
CBHW020537270326
41927CB00006B/625